PREFACE
Gardening for Life - No Money Required -
Has the potential to feed the world

Homemade "Black Gold" Compost

Planting in compost

Enjoy eating healthy

The secret lies in creating very healthy soils that work for you -
Not you working the soils

OUR PURPOSE: We want this book to provide you with simple, easy to do, low cost food gardening methods. This book is written in plain, to the point, simple photographic language. Our overall goal for this book is to motivate people to enjoy growing and eating organic homegrown foods that bring them HEALTH, HEALING, and HOPE. This is a _how-to-grow_ humanitarian food gardening effort.

Wayne Burleson is an International Food Gardener from Montana, USA. He and his wife, Connie, have traveled to eight third world countries, teaching people exciting new ways to grow food at no costs. They continue to search the world for sustainable growing methods and helping people in need. It is with great pleasure that they share their findings with you.

Their long-term goal is to network with you, sharing successes, mistakes, discoveries of new ideas. We encourage anyone and everyone to GO WILD! creating new ways to feed the world. Please post on their blog site at www.newwaystofeedtheworld.blogspot.com

Gardening for Life - No Money Required -

By Wayne H. and Connie J. Burleson

Text and Photos Copyright © 2013 by Wayne H. Burleson

All Rights Reserved, including the right of reproduction in whole or in part in any form without written permission
Designs, Illustrations and Photos by Wayne H. Burleson
Copy Editing by TJ Wierenga and Denis Paez
Printed by Midland Printing, Billings, Montana

1. Vegetable Gardening.   2. Composting.   3. Raised Bed Gardening   I. Title

ISBN 978-0-9892863-0-5

Breaking Free From Your "Yeah, buts" *Gardening for Life Workshops*

You want to start growing your own food, but in the back of your mind, from past experiences, or your neighbors telling you, "You cannot do that here!", or just from a lack of 'how-to' knowledge, you are packing around a whole list of "Yeah, buts". Yeah buts are all those reasons that stop you from trying something new. After traveling the world, teaching people that tiny gardens can return a huge portion of sustainability by providing a dependable source of food, Wayne and Connie Burleson present to you in their highly illustrated book *Gardening for Life - No Money Required*, - new ideas that will help you to bust out of your "Yeah, buts" and BECOME UNSTUCK!

What are your "Yeah, buts"?
- Poor soil
- Too cold or too hot
- Strong winds
- Deer
- Bugs (grasshoppers)
- No time
- Hard work
- Takes too long
- Inconvenience
- No water
- No $$ Too expensive
- Weeds, weeds, weeds
- Slugs, voles, birds, rabbits
- Short growing season
- No land
- Bad childhood memories
- Past failures
- Stolen produce

Rut busting your "Yeah buts" with these "Yahoos"!
- Build small gardens with big yields
- Bucket gardening in limited space
- No work gardening super easy ideas
- Multiple purpose hoop-houses extend seasons
- Connecting back to nature through gardening
- Grow your own healthy, life-healing foods
- You cannot eat a cell phone
- Eating fresh and teaching others
- Convenient "Meals on Wheels"
- Homemade compost saves big money
- Backdoor gardens save time
- Block planting saves work & energy
- Dirty hands are stress releasers
- How to build self-regenerating soils
- New "Timeless Knowledge" with less work
- Best tasting foods - eat year-round
- Improved food security
- No money required ... GO WILD!

CONTACT INFORMATION for questions and *Gardening for Life - No Money Required ... Talks and Workshops*
Travel blog http://www.newwaystofeedtheworld.blogspot.com

Wayne and Connie Burleson
Gardening for Life
332 N Stillwater Rd
Absarokee, Montana 59001

Email rutbuster@montana.net
Phone 406-328-6808

"*Gardening for Life - No Money Required*" can produce healthy & healing foods.

You can do this, too!

Our Montana Garden

Table of Contents
Gardening for Life - **No Money Required** -

INTRODUCTION	1
What is "gardening for life" all about?	1
Why these gardens work so well	2
Where and how these ideas got started	3
The needs from a world's view point	4
The needs from a United States view point	5
Learning from around the world	6-15
How these food gardens are self-sustaining at no costs	16

Chapter 1. Why Grow Your Own Food?	17
E. A. R. T. H. Why grow your own food	17-21
SAD Standard American Diet Verses a healthy diet	22

Chapter 2. Local Resource Treasure Hunting	23
Go on a treasure hunt	24
Discovering free gardening materials	25
Locating and stockpiling soil building organic material	26
Warning about contaminated organic matter	27
Save your own seeds	28

Chapter 3. Simple Ways to Make Your Own Healthy Soils	29
How to make good compost	30
Compost pile building step-by-step process	31
Making compost in Montana	32
How to keep soils healthy	33
How to make 12-day compost in a bucket	34
Why grow vegetables in soil based compost	36
How to test compost	37
The real scoop on compost	38

Chapter 4. Garden Plans, Designs & Appliances	39
Examples of small garden designs	40
African style portable food garden	41
Step by step kitchen garden preparation	42
How to construct an African style garden at no cost	43
Permanent garden bed design	44
Bucket garden test	45
Bucket gardening construction	46
How to build a long box garden	47
A wagon hoe garden	49
Instant kitchen gardens	50
Pest protection ideas	51
Inexpensive portable greenhouse	52
Double hoop-house	53
Pot hole garden	54
Succession planting in blocks	55
Ruth Stout gardening without work	56
No cost gardens that work for you, not you working for them	57

Chapter 5. Planting Guidelines for Maximum Production	59
Precision planting your seeds	60
Shotgun seeding	61
Fall seeding for spring eating	63
Backwards planning your gardens	64
What to plant and not to plant	65
African test garden before and after photos	66
Plant in multiple garden locations - don't put all your eggs in one basket	67
Double cucumbers photo	68

Chapter 6. Garden Care	69
Precision watering ideas for gardens	70
How to use compost	71
Feed the soil and it will feed you	72
Making and using compost tea	73
The best totally organic weed killer in the world	74
Biological principles that make your garden better	76
Don't weed - cultivate	80
Why mulch your soils	81
Celebrating your mistakes	82
Garden care checklist	83
How to speed up seed germination	85
No water no problem, not home no worries	86
Helpful watering ideas	87
Mr. Brite's do and don't list	88

Chapter 7. Year-round Growing Ideas - Even in Cold Weather	89
Freezing weather gardening	90
Cold weather hoop-houses let us eat for 9 months	91
Multiple purpose hoop-houses with worm cartoon	92
Wintertime carrots dug mid-February	93
Growing microgreens year-round	94
Steps to grow your own microgreens	95
Wintertime growing vegetables with two shower-stall doors	96

Chapter 8. GO WILD! Ideas - where we get to break the rules	97
What does GO WILD mean	98
Lazarus tomato trick	100
Deep mulching a green solution	101
How to kick start your garden beds	102
Build a worm cafe	103
Create worm farms right in your garden beds	104

Chapter 9. Teaching and Sharing with Others - how to grow their own health	105
Teach them young	106
Teach and tell others	107
Young people are the future	108
Fun ideas for teaching kids about gardening	109
Salmon River Pumpkin story	111
One big smile and one big potato	112
Gone wild growing healthy plants without money	113

Gardening for Life - No Money Required -

INTRODUCTION
What is "*Gardening for Life - No Money Required*" all about?

Gardening for Life is an illustrated food growing manual. It gives you simple step-by-step instructions that will enable you to produce large harvests of healthy foods from very small gardens. It teaches you some exciting <u>new ways</u> to grow food even at no costs to you, even when your daily life is too busy, even when you have no land or a limited water supply.

This garden manual will show you how to make better use of local resources currently being wasted, and how to select and test organic materials that are safe to use.

Look closely at the cover of this book. These vegetables are not store bought. They thrived in highly diverse homemade compost (some growing in 100% super healthy, soil-based compost), which is the foundation of this book: **healthy soils produce healthy foods and when you eat these foods, you become healthy!**

Gardening for Life is also about releasing you from the industrial systems of store bought foods that are altered by man. We hope that by reading this book, you become more aware of the positive effects of eating good food - and the negative effects when the quality of our food is poor. We have seen the effects that superior foods have on our own health and healing. This is one of the reasons we are so passionate about leading you toward eating more foods in God's original containers.

Gardening for Life also puts the FUN back into growing your own food by using these simple, work less, grow more, very fast growing methods. You can free yourself from the need to buy soil amendments, petroleum products, harsh chemicals, manufactured fertilizers, expensive tools, motorized equipment, even seeds and other things you think you need to buy!

Wayne & Connie Burleson enjoying the fruits of their harvest. They have traveled the world searching for and discovering new ways to feed the world.

This book will show you exactly how to put locally found treasures (clean organic matter) to work for you to provide some of the best tasting and healthiest foods that you can grow. It encourages you to put health foods back into your life and perhaps even wards off diet-related diseases such as cancer, strokes, diabetes, Alzheimer's, obesity, osteoporosis and other unwanted life-shortening ailments.

To top this all off, *Gardening for Life* could and does save lives. We all should know that if we can just eat healthy foods, we become healthier. Socrates said, "**Let Food Be Thy Medicine**"

Think before you put the next bite of food in your mouth. Ask yourself, do I know where this food was grown and what is in it? Who grew it? How many miles did it travel? How long has it been sitting in storage, sitting in a truck, or sitting on a shelf? How many nutrients are left in it? (Several vegetables held in storage lose 50 to 70% of vitamin C through oxidation.) Many of these questions are very hard to answer. We have succumbed to our local culture of just doing and buying what is convenient or eating what is handy, cheap and fast. These are not healthy practices and result in diet related health problems.

Why These Gardens Work So Well

The most important factors are that these high producing small gardens take less water, less work, and are very easy to take care of and they can be crammed full of different kinds of vegetables and once harvested, you immediately plant again. When garden soils are made with high quality compost, the plants are very healthy with a natural ability to resist many insect and disease problems. But more important than these qualities is that when looked at holistically, these gardens can become a helpful solution to our nutritionally depleted diets and to many of the world's hunger problems.

Note, a small cup waters each square

Making compost can be fun and it's free!

This compost made from garbage was placed in this 4' by 4' raised bed

Our first garden built with no money
These vegetables are growing in
100% homemade compost
They grew 100 pounds (45 Kilos) food

Gardening for Life has the potential to feed the world

Where and how did this idea - *"Gardening for Life"* - get started?

This whole effort on food gardening started with a USAID's Farmer-to-Farmer project in South Africa. My wife Connie and I were on a volunteer assignment to teach African organic farmers something called 'Best Management Practices'. We failed! Why? Because most of our modern American ideas that we brought with us cost money. Many of these people do not have money. In addition, if food was not protected near extreme poverty areas such as Cape Town's "Informal Settlements" where thousands of people are crammed together hoping for a better life, the food would be stolen. We learned a valuable lesson that modern agriculture ideas will not work under these conditions.

African Informal Settlement

We returned home to the USA discouraged but not giving up. We immediately started to search for better ways to help people that have very real needs. Quoted from an article published in the October 15, 2009 "New Times Newspaper" in Kigali, Rwanda, "The Food and Agriculture Organization and the World Food Program said 1.02 billion people – about a 100 million people more than last year – are undernourished, the highest number in 4 decades." The need is huge ... to find, test, and teach new simple, sustainable ways to feed the world, which in itself stimulates us to do more searching.

Our investigations led to Square Foot Gardening information developed by a retired engineer, Mel Bartholomew. We also studied bio-intensive sustainable agriculture methods, permaculture, hundreds of Google search engines articles, and I purchased many good books on gardening (over 150 and still counting). Our home experimentation with compost led us to testing new growing ideas that we called the African Test Garden, one entire garden constructed without money. We planted it with different vegetables growing in 100% compost. To our pleasant surprise, this vegetable garden out-performed those growing in our store-bought soils with commercial fertilizers. This book covers what we have learned about growing healthy food from around the world. We are also experimenting with many new ideas that we now call "*GO WILD! Gardening*", where we break conventional gardening rules. Now, things are exciting as we discover, explore, test and eat new homegrown "Fast Food". You can also enjoy learning how easy it is to build, grow, and eat from a g*arden designed to better your life.*
Our travel blog site is dedicated to sharing and helping others at http://www.newwaystofeedtheworld.blogspot.com

Smart phones & tablets can scan this code

THE NEEDS from a World's View Point

About 25,000 people die every day of hunger or hunger-related causes, according to the United Nations. Unfortunately, children die most often. In addition, it has been said that in third world countries, the populations of some of the larger cities will double in the next 10 to 15 years. In many of these areas, the people are crowded into slum areas, with little or no easy access to healthy food. This means that the need is not only great in rural communities, but also that urban food security will need to be addressed.

What is causing this?
This is a complex situation caused by several factors: global climate change, cycles of droughts, ever increasing populations, rising food prices, over-farmed soils, chemical-dependent soils, poor transportation, conflict, war, and in some locations – a simple lack of knowing how to grow food.

Solution:
We are excited to present to you:
*"Gardening for Life -
No Money Required" - as* an
illustrated food-growing manual,

This is a dump we visited in Ethiopia. My wife and I watched as these people were eating what they could dig out of all this garbage. This broke our hearts.

with hands-on illustrating photos to train trainers to teach these methods.

These teaching materials are detailed yet simple, covering how to make good compost, small garden designs, plant spacing, the important basic care of plants, plus harvesting and seed saving tips. We call these gardening methods, *Gardening for Life,* as they have the capability of continually producing healthy food like a well functioning organic food factory - even in many locations year-around food production without rain. **These are life saving ideas!**

Gardening for Life **is discovering new ways to feed the world.**

THE NEEDS from a United States View Point

Unhealthy eating and physical inactivity are the leading causes of death in the USA. According to the US Department of Health and Human Services, unhealthy eating and inactivity cause 310,000 - 580,000 deaths every year from diet related diseases. It is estimated that 40 to 50% of the adults in the United States are "at risk" of developing high blood pressure. Untreated hypertension can lead to stroke, kidney failure, heart attack, heart failure.

Almost two-thirds (61%) of American adults are overweight or obese.
It is hard to say how much health care money is spent trying to correct unhealthy eating problems in the US, but according to the Centers for Disease Control and Prevention, three-quarters of healthcare spending goes to treat "preventable chronic diseases." Treatment for obesity alone runs an annual tab of **$147 billion**, and that does not figure in diabetes (**$116 billion**) or cardiovascular disease.

What is holding us back from eating better? I would say the underlying culprit is laziness to convenience caused by a new culture of being too busy. It is so easy to buy what we are accustomed to eating. A 'grab and go' style of eating is just so much faster and more convenient.

So how can we change? Well, we need to invent simple ways that help us to curb our old ways, and get excited about preventing some of our sorry eating habits. In addition, we should go back to the kitchen and invest some healthy time on learning how to cook so that we can eat better. Time well spent.

Just think, how many children in the United States graduate school actually knowing how to cook a few complete meals from scratch? A very low number.

Solution - Reacquaint yourself with growing, cooking and eating real whole foods for health and healing. It is worth more than the time and effort it takes. It can be a matter of life or death. You choose.

Processed Foods

We are what we eat. You choose

Whole Natural Foods

Gardening for Life - No Money Required - has the potential to feed the world.
And here is the proof

We have had the privilege of teaching food gardening projects in eight third world countries. The following is what we have learned.

Learning from Others

Wuppertal, South Africa

Here I am, a greenhorn Montana fellow, visiting about food security with an astute South African farmer.

This was our first country in attempts to teach Africans some better ways to farm and to grow organic food gardens.

What we learned from him was that our nifty modern ideas cost way too much money. These people were just fine living like people did hundreds of years ago, mostly farming without money.

This remote African farmer was hard working, proud of his small crops, super friendly and a good farmer. Like me, he also liked John Wayne, which aided our good communication.

He helped us to see that his simple ways and wise use of local resources worked just fine.

We also learned that millions of African people lived in Informal Settlements where finding enough healthy food is a daily struggle.

Innovative Marketing in Malawi

Lobe, Malawi, Africa

Our second trip to Africa was more productive.

By this time we had learned the single most important lesson, if you want to grow a great food garden, you must start with great soils. You can make these soils yourself from unused (often discarded), untreated, "clean" organic material. In this context, the word "clean" means "not contaminated by man".

Here we are in a remote village teaching the basic ways of making soil-based compost, using only local resources that cost no money. We made improvements in our teaching methods to help others help themselves. TEACHING TRUE SUSTAINABILITY!

This compost pile has alternating layers of dry grass, green grass, dry livestock manure, top soil, and a sprinkling of wood ash.

Homemade Potato Chips - an innovative marketing idea

The Malawians taught us a great way to market and sell potatoes.

They used a large sheet of metal with a pit in the middle, placed on a rocked-up wood firebox. The metal pit in the center has hot oil in it for cooking potatoes.

The marketers just chopped up the potatoes, slid them into the pit of hot oil, and cooked them until done. You paid a small fee and they slid your potatoes over in front of you. You sterilized a fork by sticking it into the hot oil, then stabbed your potatoes and dipped them into the pile of salt. Yummy, very hot potato chips!

They were great tasting and again almost no money was spent to sell their crop of potatoes - selling a product - value added.

Firewood Galore

Lobe, Malawi, Africa

All over Africa, we observed how creative people could survive living on limited resources.

Here is a skinny-wheeled bicycle loaded down with about three wheelbarrow loads of cooking firewood. This is an amazing feat of African ingenuity and another example of how to make great use of limited resources.

I am not sure how they ever stacked this heavy load on the bicycle or what would happen if it tipped over.

Truly amazing!

Below these Malawians located several old wasted clay bricks to make this very handy sit-down kitchen garden. You can sit on the bricks while tending the garden, which is easy on the back muscles. Again, no money required. These small gardens were watered with recycled wash water.

Success in Rwanda, Africa

Same garden 60 days later

Rusera, Rwanda, Africa

We tested several demonstration kitchen gardens in this small African community. This trip proved that we could use recycled wash water as the only source of garden water. This idea made a big difference to the Rwandans in growing their own food, as they have to hand-carry water a very long distance during the dry season.

We also planted a very rare, hard-skinned, winter squash that grew profusely in Rwanda's climate. We believe this plant comes from American Indian cultivation in the warm canyons of the Snake and Salmon rivers in Idaho, USA with a very hot climate. The squash seeds are not yet for sale, which means this vine crop can only be grown from saved seeds. We taught them how to save their own seeds with this squash plant, which is a vital example of why it is so important to learn good seed-saving techniques.

The real test of going from seed to stomach passed during our three-month stay in Rwanda. The African folks loved the taste of this squash plant. Pastor Cosmos pictured here is enjoying eating his Salmon River Pumpkins. This was a trip coordinated with the Saddleback Church P.E.A.C.E. Plan.

Children learning skills that help their families

Malawi, Africa

These children passed us each day, carrying water, singing and helping each other while providing a valuable service for their families. We all could learn a valuable lesson from these kinds of family survival activities.

Ethiopia, Africa

Third world countries taught us that they are raising their children to help with the sustainability skills of their families, which is a good thing. These kids are learning how to contribute and understand what it takes to live life from the earth, which gives them worthy purpose, just as the farms of yesterday did.

Modern times have moved people away from these simple survival skills, making them dependent upon outside resources. *Gardening for Life* brings us back to the land, where we become better caretakers of the soil. We learn that healthy soils grow healthy food to make us healthy. Let us all teach our kids these life giving skills and how to take better care of themselves, independent from outside resources.

One Whopper Radish

Shone, Ethiopia, Africa

Look what happens when you plant one radish seed in good soil. One huge radish grows, which could feed a whole family.

After seeing this photo, we did some internet checking and found out that we should also eat the tops of the radish, as the leaves have 3 times the nutrients as the roots.

My wife is not a big radish fan. However, I made her a complete salad out of one radish with a bit of salad dressing and she loved it.

We are still learning that growing healthy food is just **one** of the important parts, but just as important is learning new healthier ways to eat the foods that we grow.

Shone, Ethiopia, Africa

Several of these gardens were planted in Ethiopia. Everywhere we went the people would come and watch for a while. Then they would tell us that they also wanted a food garden.

People living without much are very interested in learning new methods to grow their own food. They become excited as they watch simple 'no cost' gardens being constructed. The world needs to grow healthy food using these kinds of simple, self-sustaining, life giving gardens built without petroleum products.

This is what *Gardening for Life* is all about. The first and most important step is to learn how to build super healthy soils that work for you, instead of you working the soils. The health factors are inside these soils.

'Ya Mon' Jamaica

Near Port Maria, Jamaica

Here we are celebrating Gardening Power while constructing terrace gardens on the island of Jamaica.

This is truly **Gardening for Life** in the tropics where it can rain 3 inches in one hour. We experienced unbelievable erosion as we watched a powerful rainfall event move asphalt pavement from a hillside roadway.

By unplanned success, we learned a valuable lesson when building these hillside garden beds lined with bamboo logs. We mulched the heavy, native clay soils with partially composted chicken manure mixed in with woodchips. The woodchips held these soils together and stopped the erosion from these recurring torrential rainfalls. We considered this successful idea a happy accident that prevents "raindrop-splash erosion".

St Mart, Jamaica

An organization called American Caribbean Experience (ACE) is located on the North side of the tropical island of Jamaica. They operate a missionary style work/vacation experience for groups of volunteers.

ACE has plans to teach every child in their sponsorship program as well as all the student in their garden clubs how to **'grow what they eat and eat what they grow'** while teaching them to grow their own vegetables. They are going to build garden boxes at each of their sponsorship students' homes.

We also learned another valuable lesson about eating right from this trip. Homegrown food is much healthier then consuming highly processed foods. How can I say this? After our trip to Jamaica, I had my blood tested by the USA Veterans Administration. My cholesterol levels were much improved. A prime example of **'we are what we eat'**. You see, we were eating more fresh vegetables.

Before and after success in Jamaica in only 60 days

Tijuana, Mexico

On a mission trip with the Saddleback Church form Southern California, we had the privilege to test a great new way to feed the world. *Sunflower Microgreens* are simply black oil sunflower seeds growing into the first two-leaf stage. The leaves contain 20% protein plus other living nutrients.

These kids in an orphanage just loved eating them. We needed a whole boatload of these great tasting microgreens, which can be grown in a pot in only 8 days. Now this is real and healthy still alive - **'Fast Food'**.

Tijuana neighborhoods

Microgreens - an 8 day crop

Teaching bucket gardening in Tijuana, Mexico

If you can grow it in a regular garden, you can also grow it in a bucket garden. Here we are demonstrating planting food gardens in 5-gallon buckets at the Tijuana Orphanage.

We learned in Tijuana that the garden stores did not have any vegetable transplant for sale. They only had starter pots of flowers and shrubbery even at the large Garden Centers like Lowe's. This seemed strange as over one million people live here and in a two-hour search, we could not locate any tomato or pepper plants.

HEY COME ON, this is MEXICO!

Leon, Nicaragua

These folks have just completed a training session growing their own food in a box garden.

We learned a valuable lesson when teaching in a foreign country. First, teach a short translated session to the class, and then let the students teach the same information in their own words back to the whole class. In this way, the class gains a greater understanding of the individual session. They hear it twice. In addition, it is a huge help to have someone continue your teaching efforts after you leave.

This is exactly what happened here with Peace Corps volunteer Britney Stanley (standing next to me).

Students are holding certificates form class completion. This trip was coordinated with the Partners for America, Farmer to Farmer Program.

Teaching an old dog, new tricks

Barrio Pio XII, Nicaragua

Gardening for Life is as much an art as it is a science. Here I am learning from an expert Nicaraguan farmer the art of plant grafting with a very sharp knife.

He is teaching me how to graft a new and better bud onto the healthy root system of another plant. If done correctly, you will end up with a very healthy flowering system that is much more productive than either the original plant or the newly grafted bud.

Another important lesson: do not think you have it all figured out and become closed-minded. Everywhere we go we continue to learn more new, innovative techniques that make *Gardening for Life* an exciting growing experience. Pun intended.

Praying over a 'Long Box' garden

Nhangau, Mozambique, Africa

What we learned in Mozambique is that if you give a person a seed, the plant may stop growing one day. But if you teach a person to save seeds, new plants will continue to grow and you can change a whole community.

That is what happened here in this very small African village. Our excellent translator, Mr. Vengai Rufu Chikono (in the white hat in the photo to the left), started a community organic agriculture school, teaching his whole community new and simple ways to grow their own food.

We toured the local market and found out that almost all the food was imported. How sad. This means that the local market owners are just the middlemen selling someone else's produce, probably for a very small profit. However, that will change as Mr. Rufu teaches his community *Gardening for Life* - new ways to grow and sell their own local crops.

Nhangau, Mozambique, Africa

Finally after all these trips to third world countries, we can say that we have learned better ways to teach people sustainable methods to grow healthy food right in their own backyards without having to go and buy a bunch of stuff. It is all about how to make your own healthy soils to grow healthy plants to harvest free sunlight energy.

Here we are celebrating building a compost pile. If you look closely into the faces of these folks, you can see and feel their enthusiasm.

After all this traveling and learning, we have a different worldview of what is going on around the world. Factually, we can say:

***Gardening for Life* - No Money Required - has the potential to feed the world**

Growing "Black Gold" compost

How these food gardens are self-sustaining at no cost

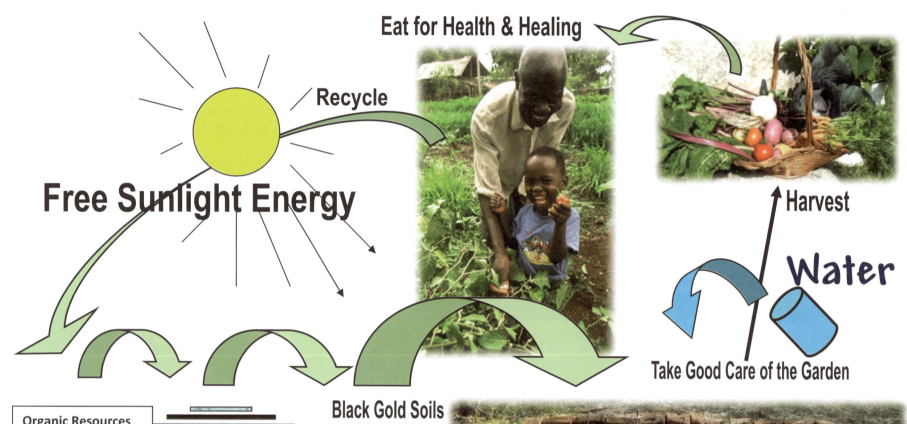

Chapter 1. Why Grow Your Own Food?

E.A.R.T.H.   We are what we eat!

We should grow our own food to:

E = To capture free sunlight **ENERGY**

A = To help **ALLEVIATE** world hunger and **ADDRESS** food security problems at home

R = To **REACQUAINT** ourselves with the **REAL** taste of **REAL** food by pulling healthy nutrients from 'black gold' soils with a fresh taste that can't be bought
   In addition, to **RECYCLE** your organic waste to be made into compost which rejuvenates your soils

T = To save **TIME**, **TRANSPORTATION** and money
Fresh vegetables are full of life-healing enzymes

H = To grow your own **HEALTH**
Healthy soils produce vigorous plants that when eaten fresh, **help you** become fit

Besides, something unexplainable happens when placing your **hands** in rich, dark-colored soils teeming with life, which are growing healthy **vegetables**

E = To capture free sunlight ENERGY

If you do not capture energy from the sun, you lose it. A good rule to follow is to not let sunlight hit bare soils, as you have just wasted some good growing power. Instead, plant your gardens with wall-to-wall vegetables to capture this free energy. After harvest, stir in some new compost and replant immediately - now the earth, sun, and water are working very efficiently for you.

Open your hand and place it palm up, directly toward full sunlight. The warming sensation you feel is free sunlight energy. This is the driving force behind the amazing miracle of photosynthesis, which is the natural process that feeds the world. No sunlight, no food.

Think of green leaves as solar panels. Some plants have the ability to keep their leaves turned directly toward the sun, following it as it moves across the sky. What an amazing miracle of nature. Just ask yourself, am I efficiently capturing free energy? Or am I giving my hard earned dollars to industrialized marketing companies to feed my family?

Look what free sunlight energy can grow
Gardening for Life **has the potential to feed the world**

Have you ever considered where plants really get their food? Most of us will say plants pull food from the soils. This is partly right. However, each leaf has the built-in ability to absorb this free energy, and convert it into a sugary substance (carbohydrates) inside the leaves. This sticky liquid moves around inside the plant feeding the plant cells so that the plant may grow. So the truth of the matter is - plants don't get their entire food for growth from the soil, instead they make their own food right inside all the green parts of the plant using certain organic minerals from the soils and free sunlight energy.

A = To ALLEVIATE world hunger problems and ADDRESS food security

We doubt if food will become any cheaper in the future. We have heard of rumors that milk prices may double. Eroded farm soils make for higher and higher farming expenditures, and up goes the cost of food. Our local grocery store owner said that if his delivery trucks stop running, his store will empty in 3 days. Now that is a scary situation. This is probably connected to the national movement to become more concerned about our food security. I do not want to think about what the public would do if there ever became widespread food shortages. This is how wars start.

About 1/6 of the world's population struggles with inadequate food supply; however, that number would change if more and more people would adopt some of the simple homegrown food growing ideas presented in this book.

R = To **REACQUAINT** yourself with the **REAL** taste of **REAL** food.

Most people know that highly processed food does not have that great taste (or all the nutrients) when compared to eating directly out of a garden. Fresh picked tomatoes form the garden, popped straight into your mouth, have that real tomato taste you just cannot find from the store bought version.

What is going on here? After picking vegetables, they continue to breathe. This process, called respiration, breaks down stored organic materials, such as carbohydrates, proteins and fats, and leads to loss of food value, flavor and nutrients. According to the Department of Food Science and Technology, University of California - Davis, vitamin C degrades rapidly after harvest, and depending on which vegetable you are harvesting. As much as 77% of the nutrient in green beans may be lost in 7 days even in cool storage. Along with B vitamins thiamin, B6 and riboflavin are sensitive to heat and light, resulting in additional nutrient loss.

We once had a tomato-tasting contest to see which variety of tomato had the best flavor. We tested yellow Brandy Wine, Sweet 100, Cherokee Purple, Early Girl, Celebrity plus a few others. One tomato slipped into this contest was purchased from our local grocery store labeled "Grown in Mexico". Guess which one was on the bottom of this taste test? Yep, the Mexico tomato, which tasted a bit like cardboard, lacking that great tomato taste. Obviously, this tomato was genetically bred for a long shelf life. The winner was the Cherokee Purple tomato, selected as the best tasting.

A garden club tomato tasting session

Why grow your own food? To eat fresh great tasting foods still alive and full of health & life-healing enzymes. In addition, **RECYCLE** your own organic waste to grow more food. This is the foundation of *Gardening for Life*. If you grow in soils made of homemade-composted, highly diverse organic matter, with a soft deep soil structure to help promote fast root development, you will have a green thumb. One great indicator of healthy soils is the number of earthworms present. Healthy worms, healthy soil, healthy food. It is nice to count 60 earthworms per shovel-full in our garden soils, a great indicator of vigor.

T = To Save **TIME**, **TRANSPORTATION**, and money

Time is something that most of us take for granted. It just keeps moving. A home garden saves time by allowing you to just slip out the door, grab some vegetables and fruit from the vines, mix it into a salad and there you go. You save all that time not needing to go to the store, park your car, find a cart, shop, pay for it and then drive home.

Hint: build your gardens as close to your home doors as you can. The handier you make food gardens, the more you will use them. Consider planting in planters just outside your kitchen door with several different kinds of salad-making greens. Grab a bowl, head out the door, cut the larger outer leaves and leave the center of the plant (hearts) to grow more leaves that are delicious for next time. This is called 'cut and grow' gardening.

The truth be known, many of us cannot afford to buy expensive organic vegetables. So grow organic yourself. I once thought that I spend way too much time working in our garden. However, a good friend straightened me out fast. She said, "Wayne, don't you realize that what you're growing, you cannot buy! It's super fresh and organic." She was right.

H = To grow your own **HEALTH**   Healthy soils produce healthy plants that when eaten fresh, help you become healthy.

The idea of growing your own food is becoming a really big deal. There are billions of people around the world that are in need of simple food growing ideas, connected to sustainable, high production gardens.

It is important to know, **we are what we eat**. Hippocrates, an ancient Greek physician way back in 460 BC, stated, "Let your food be your medicine, and let your medicine be your food." He also stated, "Medicine should do no harm." In today's world, most of our foods are processed, pre-packaged, preserved for storage with chemicals, and then shipped across the country. All of this is an expensive effort and has a deteriorating effect on the foods we eat. It is just the way it is.

However, we can choose not to participate in always eating processed foods. The following illustration is the basic reason to grow your own food. I know of no other way to know what is in your food than to grow it yourself.

HEALTHY SOILS   =   HEALTHY FOOD   =   HEALTHY PEOPLE

After eating a complete meal from our outside garden, my wife and I pushed our dishes back and reflected on our feelings. Something about growing your own food gives you a feeling of accomplishment; digging in all that nice looking black soil just feels good. We also benefit from the exercise of caring for a garden, lots of bending.

In review, growing your own food is good for our hearts, it keeps money in our pockets, and fresh picked is better than you can buy anywhere around this small rural community. In addition, did I mention it was wintertime mid-December, and we live at 4,400 feet in elevation in Montana? Just the other night it had been seven below zero and I harvested mixed lettuce, spinach, and carrots that along with stored red potatoes and winter squash, became our complete meal.

This photo below shows the exact mixed lettuce that we pick from our outside garden. Protected by a small, portable hoop-house (a hoop structure constructed out of stiff wire with plastic sheeting attached), this is great looking lettuce and better yet, great tasting organic leaves. Just the sight of all those different kinds of beautiful vegetables leaves is mouth watering. So do not let a bunch of your 'Yeah, Buts' (all your reasons not to grow) get in the way of growing your own super healthy food.

These good-looking mixed green leaves were growing under that snow-covered hoop house - mid-December in a Montana winter

Standard American Diet   Versus a   Healthy Diet

This S.A.D. diet is sad because it causes diet related health problems. Simply put, too many carbohydrates and too much sugar in this eating style. The root problem is **overeating, yet undernourished.**

If you want health, chose this eating style. Half your plate should be fruit and vegetables.

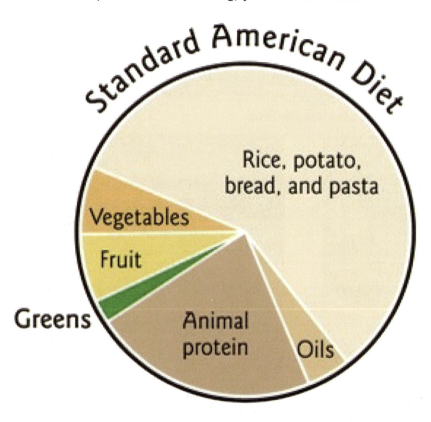

We are what we eat.   Just remember, we do have a choice.

Learning to grow your own health by using some of the fast, easy, low cost methods explained in this book could make your life better. If you want to stay or get healthy and fit, gardening for your life is a good idea.

Chapter 2. Local Resource Treasure Hunting

Free Leaf Mold

Kitchen Waste

Go On a Treasure Hunt Searching for Hidden Resources

Can you name all the hidden resources in this photo?

The above photo looks like a junk pile. However, all of the necessary materials are available here for you to build a food factory garden and make a compost pile.

Most locations anywhere in the world have these resources going to waste, just waiting for someone to make good use of them.

Take a sack and walk the neighborhood collecting wasted organic matter

Drying livestock manure can be used for gardens or burned as fuel for cooking. (Photo taken in Ethiopia)

What to look for:
Dry grass and leaves, kitchen waste, potato peels, old bananas, black top soil, green grass and leaves, old livestock manure, and especially dark colored soils under old piles of rubbish.

Where to Look:
Anywhere and everywhere. Look at local farmers markets which have spoiled and damaged vegetables, under the vegetation along fence lines, cow barns, chicken coops, rabbit hutches, under trees and shrubs. Even pay kids to walk the roads and fields and bring you sacks of dry grass and green grass, old dry livestock manure, and vegetable waste.

How to use it:
Collect all these organic materials and place them in piles next to your gardens. Make compost piles out of all the different organic matter by layering them like a big sandwich, so that it will heat up and make good compost to add to your garden soils.

Discovering Free Gardening Material

Gardening for Life is all about being able to grow food without spending money. So here is where your powers of resourcefulness come into play. I'm always scouting and on the lookout for discarded garden building materials. You can drive by construction sites, survey dumps, look for rock outcroppings, be on the watch anywhere and everywhere you go. Most times when you ask to help someone clean up after some home construction project they will gladly let you haul their waste material away. You have just saved the world by using some wasted resources headed for the landfill.

Don't forget to look into your own back yard or drive up and down alleyways. You may find some great, even attractive recyclable treasures this way.

The following is a short list of goodies to look for:
- Old discarded lumber for sides of a garden box
- Large flat or rounded rocks to use for edges of garden beds
- Cement or clay bricks that nobody wants any longer for stacking to build garden beds
- Clean cardboard boxes to lay down in order to choke out weeds and grass growth
- Black and white newspaper to lay on the soils for weed control
- Non-treated logs for sides of gardens
- Bamboo - very handy for staked plants or building sides of boxes
- Wood sticks for stakes
- Thick plastic sheets (like pond liner) for walkways between beds to stop the spread of grass
- Be creative and come up with your own unique designs

The photo here depicts what we have named a 'Long Box' Garden. It's 4 feet wide and 40 feet long, and filled completely with homemade compost. There was no plowing or tilling under this box, as I simply placed layers of cardboard and newspapers on the existing soil surface before stacking on the compost. This stopped all the unwanted vegetation from growing up through the garden bed, and worked great.

The production out of this one garden could feed a whole family. I measured 1.5 pounds of mixed lettuce from one square foot in one cutting. This one box has 160 square feet in it. That equals 240 pounds of food produced in only 40 days. If you could grow three or four crops from this 'Long Box' Garden, you could produce as much as 720 to 960 pounds of food. In theory that would feed 5 people one pound of vegetables per day for 192 days. Not bad for one garden box made completely with no cash outflow, as we do save our own seeds (see chapter 6).

Locating and stockpiling soil-building organic material

Organic matter is what is missing in most soils. Start locating, transporting and stockpiling these kinds of natural materials as indicated in the photo below. This photo was taken in Malawi, Africa as an example of wasted resources just lying around the neighborhood waiting to be used. Anywhere in the world, you can find someone wanting to get rid of these kinds of garbage items. They are not really garbage, but should be considered valuable natural resources. By doing this, people save money, recycle waste, convert garbage into valuable growing power, and slow down the pollution and the added costs to dumps. Good for the environment.

It's a good thing to have these kinds of valuable organic materials stockpiled somewhere nearby your garden area for handy use. By stockpiling ahead of time, you will greatly speed your soil building ability through the magical process of composting and mulching.

Top soil (Microorganisms)
Look for dark colored soil with organisms.
Do not use treated soils.
Stay away from man-made contamination.

Livestock Manures (Organic Fertilizers)
Cow
Sheep
Horse
Donkey
Chicken
No pig, dog or cat

Green Material (Nitrogen --- Proteins)
Green grass
Flowers
Green roots with soil
Kitchen waste
Egg shells
No meats, oil, or cooked foods

Dry Material (Carbon --- carbohydrates & sugars)
Dry grass
Straw, Hay
Dry leaves
Saw dust, cardboard
Paper waste (shredded)
Yard sweepings
Wood chips

Clean wood ash (Minerals)
A small amount of non-treated burned wood ash. Do not use treated wood.

Warning about contaminated organic matter

A good friend of ours told us about a huge mistake she made in her garden beds. As the story goes, someone offered her pickup loads of old composted livestock manure. She, being a master gardener, asked all the right questions to the farmer about how this manure was made and what ingredients the cows were fed, just to make sure she could stay as close to being natural and organic as she could.

She received the composted manure, hand dug it into her garden beds, planted in it and everything started to grow gangbusters (fast and looking good). Then certain plants like her peas stopped growing and the tips of leaves started to curled inward. She thought this must be some sort of mosaic virus or other plant disease. Later while attending a workshop she just happened to have some samples of the leaves with her that she showed the instructor, still thinking she had a disease problem. However, the instructor examined the leaves closely and told her, "This looks like herbicide contamination".

Sure enough, they traced back the history of the cattle grazing several years earlier and discovered that these cattle had consumed some plants that had been sprayed next to a patch of noxious weeds which had been sprayed with a restricted herbicide registered for control of this particular weed. The chemical compounds came into contact with the neighboring grass plants, and the cows consumed the grass with these chemicals on them. The chemicals passed through the cows and ended up in the livestock manure. Our friend had to go back and hand-remove all that contaminated composted soils, which she had carefully double dug into her garden beds, and replaced it with clean organic matter.

The warning here is... be careful when accepting organic material that you do not have specific knowledge about. Commercial lawn care experts have a whole host of chemicals to keep things looking neat and manicured these days. Become knowledgeable what is in your material (herbicide, heavy metals and other contaminates). If you do not know, probably do not use it.

This makes a case for questioning free compost from city dumps. There is no way to know what ingredients are in the original picked up green boxes that municipal sanitation departments haul away and compost in large piles. Be careful and think about a closed system so you know where your organic material comes from and what is in it.

I have used some of my neighbor's sheep livestock bedding and hauled it home for composting only to end up with some new weed species in and around our garden beds.

One way to combat these kinds of situations is to use more autumn leaves. I know that it takes most of the wintertime to turn leaves into soils (see the next chapter on how to do this). However, the leaves are usually higher in the tree where people don't spray as much.

Save Your Own Seeds

This is another resource that goes unused. You can even save seeds from rotten produce.

Spinach
Pick out the strong plants and let them bolt into a flower stalk and go to seed. Pull the seed stalks out of the ground and let dry. Thresh the seeds into a container.

Cucumber
Let ripen past edible stage and turn yellow. Cut lengthwise, scoop out seeds and dry.

Beets
Biennial, as it takes two years. Store roots for several months, replant to grow seeds, harvest seeds when dry.

Pumpkin
Cut ripe & mature pumpkin open. Remove seeds. Wash with water. Place on screen or cloth to dry.

Onion
Let a few plants form round flower clusters. When dry, pick and thresh the seed out.

Pepper
Let ripen to full color, no sign of disease. Remove seed off core and place on screen or cloth to dry.

Tomato
Pick ripe tomatoes from several plants. Squeeze seed out, wash and spread on cloth to dry.

Lettuce
Allow plant to bolt, to form a seed stalk. Cover to protect from birds & rain. Harvest seeds for 2 to 3 weeks. This will require repeated harvesting.

> IT IS MISERABLE FOR A FARMER TO BE OBLIGED TO BUY HIS SEEDS; TO EXCHANGE SEEDS MAY, IN SOME CASES, BE USEFUL; BUT TO BUY THEM AFTER THE FIRST YEAR IS DISREPUTABLE.

A quote from George Washington

Chapter 3. Simple Ways to Make Your Own Healthy Soils

HOW TO COMPOST

If you don't feed the soils, they won't feed you!

Mr Brite

Step 2 ... Build a compost pile close to your garden

COMPOST INSTRUCTIONS:

Repeat all layers until 1 Meter high
Add small amount of wood ash
Add water to dry layers →
Vegetable waste ------→
Thin layer old manure -→
Thin layer top soil-----→
Green grass 1' - 30 cm --→
Dry grass 1' - 30 cm -→
Dig soil bottom for air --- >

Water pile when dry

A Compost Pile

Step 1 ... Locate the following organic material Move & stack them near your gardens

Top Soil, Green Plants, Old Dry Manure, Small Amount of Wood Ash, Chop Vegetables, Egg Shells, Dry Plants

Why compost? Compost is decomposed organic matter that has turned into black colored humus referred to as *"black gold."* Compost makes excellent organic plant food. Millions of micro-organisms digest the dry grass and green grass, causing the pile to heat up. Compost does not feed the plants directly, instead it feeds the soil microbes, which in turn release insoluble minerals for the plants to feed upon (fertilizers). This amazing process makes your garden a sustainable food factory if you keep adding compost to your soils.

Step 3 ... the compost pile heats up. Turn & mix pile once per week.

Step 4 ... After about 60 days or longer when compost turns to dark colored soil, mix into garden.

Black Gold

Add compost to garden soils often

Compost Pile Building - Step-by-Step Process

Loosen soil to allow air under the pile

Add layers of course organic material to allow more air

Add 1 layer of dry (20 cm) & 1 layer of green matter (20 cm)

Add a layer of old dry livestock manure
(Use chicken, cow, and/or horse)
(No pig, dog or cat manure)

Add a thin layer of top soil

Add a small layer of clean wood ash

Add water to the whole pile

Continue to repeat all layers until pile is 3 meters tall

Lay hands on and pray. Turn pile over once per week

Making Compost in MONTANA USA (how to grow your own good quality plant food - compost)

This is how we make compost in Montana USA. Layer 6 inches (15 cm) of dry grass clippings with one inch (2 cm) of aged cow manure, wet each layer and add more layers of grass clipping and manure. Add green grass and garden waste along with some chopped-up kitchen vegetable scraps when they become available. Build the pile 2 to 3 feet (one meter) tall and as long as you need. Keep adding several different kinds of organic materials in layers. Let sit for about two weeks, then mix. Feel inside the pile for heat, the good indicator that compost is happening. Add more water if needed and turn often. The more you turn the pile, the faster it will make compost. You can just fork over the whole pile into a new row. We cheat and use the tiller.

When the compost is ready to be used, the original organic matter turns black in color (**to "black gold"**) indicating humus. Humus is a stable form of carbon that plants love and grow very fast in – it is nature's ideal plant food. The compost will have a pleasant earthy smell, holds water well, stays soft and fluffy – which resists compaction, and of course is full of high quality plant nutrients. If you use only organic materials, you can call this naturally organic compost gardening. Compost gardening has the ability to keep and hold nutrients for a long time, which creates a more sustainable system that keeps on feeding the soil. If you keep adding the compost often to your garden soils during the growing season, your plants will remain very healthy, giving them the ability to better resist diseases and insect infestations.

How to Keep Soils Healthy

The best natural process to maintain soil health is to keep the soils covered with something. Mulch, mulch, mulch.

Clean weed-free straw, leaves and/or dry grass work well as mulch and should be applied often to the top of your soil. You can pile it on 4 - 8 inches deep between individual plants.

We have found that once a garden bed is full of vegetables, that is, when the plant leaves touch each other, they form a living mulch. Other plants and weeds cannot find room to invade.

When block planting like in the below photo, be sure to fall mulch so that the soil critters are well fed throughout the fall, winter, and springtime.

Living Mulch

Down Yellow Litter, Feeds Small Soil Critters

Healthy soils are alive soils
They (the soil critters) need a roof, food, and water.
SO! "KEEP THE SOILS COVERED"

How to Make 12-day Compost in a Bucket
Fast healthy compost that looks like dry compost soup

What is Compost:
Compost is simply decomposed organic matter that has turned black or dark brown in color. Organic matter includes garden waste, kitchen scraps, old manure, dry leaves and dry grass, green plants plus other organic material. Compost happens when billions of soil micro-organisms eat (digest) organic material and changed the carbon into valuable humus. Humus is one of the world's best homemade, free plant foods.

Tools needed to make 12-day Compost:
Equipment and tools needed:
- A 5 gallon plastic bucket with handle and lid
- Or something to cover the bucket the with holes in it to allow oxygen to enter into bucket.
- Drill holes in bucket with 1/4 inch drill bit or use a large nail and hammer air holes into the bucket - top, bottom and sides.

Recipe for 12-Day Compost - a dry soup method:
- 1/3 bucket full of *finely chopped-up* dry leaves and dry grass (carbohydrate - a form of sugar for energy)
- 2/3 bucket full of *finely chopped-up* green plants (nitrogen to build protein use for muscle building)
- 1 shovel full of good top-soil, dirt around roots, and/or old compost (add as an soil organism activator)
- 1 small shovel full of old dry chicken, cow, horse or donkey manure (more nitrogen)
- Hand full of chopped-up egg shells, kitchen waste such as carrot tops, potato peelings, banana peelings, and/or weedy plants without seeds (for added biological diversity)
- Air provided by holes drilled in the bucket and occasional stirring with a hooked stick or metal rod
- Water added to keep dry material wet (keep moisture in bucket to feel like a damp sponge)
- Stir or roll bucket several times and check after 12 days
- Use decompose organic material (compost) when it changes in color and feels like black soil (Black Gold)

12 days later leaves turned into dark colored compost in same bucket

12-day Compost added as mulch under this eggplant speeds growth

How Best to Use 12-day Compost

Compost is a valuable soil additive when added as mulch around the base of plants. It is one of the best-known free, natural slow released fertilizers. Plus, it improves soil structure, aids in necessary microbial activity in the soil, and attracts beneficial earthworms. Compost can suppress several soil borne diseases and it holds nutrients in organic slow-release form, allowing for availability throughout the growing season. Compost holds moisture, which greatly helps to conserves valuable soil moisture.

Compost, being rich in plant nutrients, is used in kitchen gardens, farming, landscaping, and other horticultural projects. The compost itself is beneficial for the land in many ways, including as a soil conditioner, a fertilizer, addition of vital humus or humic acids, and as a natural pesticide for soil.

If you want to speed up growth in your garden beds, try making liquid compost tea out of 12-day compost. A simple fast way is to locate a 1-gallon glass jar, place 3 inches of compost in the bottom of the jar, fill with water, cap it off and let stand for a few days. You are turning the compost into liquid fertilizers.

We add this compost tea to bucket gardens after the plants get big and their root system have hit the bottom of the bucket. They need more plant nutrients. Just fill your watering can with this liquid tea compost and give your needy plant a drink. This should be done about once a month.

Compost tea ready to add to these transplants

Why Grow Vegetables in Soil Based Compost?

The answer to this question is '**to get healthy**'. Good compost is highly diverse, decomposed organic matter that contains rich humus with huge amounts of different chemical compounds. Soil based composts are full of living organisms such as fungi, protozoa, beneficial bacteria, arthropods, nematodes, centipedes, springtails, insect larvae, and earthworms. These kinds of soils are teaming with billions of living organisms, which is quite different from depleted, long-term used, over-processed farm soils that are becoming more and more deficient in soil nutrients. In other words, do not grow food in poor soils that are dying.

Here is the most import reason to grow your food in diverse compost. Brett Markham published a book entitled "The Mini Farming Guide to Composting" and in it, he writes, "Compost contains and preserves these micronutrients that are so important to human health." Today, many of our store-bought foods are growing in nutrient depleted farm soils.

I do know from experience that the first crops raised in newly plowed soils grow well the first few years. However, year after of year growing crops in these same soils, plants start a downward trend in quality. Modern agriculture deals in total production and profit incentive and not so much in total nutrition. Certain farms end up buying more and more petroleum-manufactured fertilizers to continue to produce crops. That is just the way it is.

If health and preventing diet-related problems is your concern, please consider making your own CLEAN, organic, highly diverse compost. Clean means you know what is inside your organic materials. We like to use tree leaves, as it is hard to treat (spray) tall trees with toxic chemicals. We also try our best to keep track of spraying programs that go on around our neighborhood.

Healthy transplants grow in homemade compost

We do not want to grow acres of vegetables; instead, we just want to grow super healthy food that provides health back to us. You can make garden vegetables look healthy by applying commercial chemicals, but are they? Some compost is totally plant-based as no soil has been added to the compost. I personally believe that is not the best idea. It is much closer to nature to grow connected to the whole earth (soil-based compost) as that is where most natural plant nutrients are formed. Meaning GO WILD and let the native soil microbes do the work for you, so that we can eat healthy without adding store bought plant nutrition. We try our best to stay natural organic.

Let nature be our teacher!

How to Test Compost

Which plant nutrients may be missing in your compost? Good question. If compost is as good as this book is saying it can be - how do we really know? One way is simply looking at your plants. Are they growing fast, do they have dark green leaves, and are the stems strong, thick and growing straight? If not, what is missing?

Simple soil tests reveal some clues. A color-coded test for nitrogen, phosphorus and potash will put you into the ballpark on which macronutrients may be missing. You can also check the pH level to know how well certain plants are able to receive these nutrients. Use the information that came with your soil tester kit to figure out if soil adjustments are necessary.

If the pH range is somewhere between 6 and 7 your soils are probably OK for growing most vegetable plants. Growing in compost is considered one of the safer pH growing mediums.

I suggest if you are unsure of how clean (free of herbicides) some ingredients are in your compost are, try a test pot with peas and/or beans. These are legume crops. Watch the leaves closely; if the outer edges roll upwards, it is an indication that you have some sort of contamination problem.

Spread some of your compost out on a sheet of white paper. If you see lots of identifiable plant parts such as stems, leaves, or wood chips in it, it has not completed the composting process. Also, take a hand full of compost and squeeze it. Good compost will fall apart into many dark colored small chunks.

Smell it. High-quality compost will not have a bad order like ammonia, but have that great, fresh, sweet-earth smell, which we say is heavenly.

Curled leaves in peas could indicate that some form of herbicide contamination is present in the soils or compost. Have these soils tested by professionals.

The Real Scoop on Compost

The world's BEST natural fertilizer is highly diverse, soil-based compost. How can I say this when some of the conventional-thinking literature says 'compost should not be considered a fertilizer'? Well, just look at the photo below. This is a field of native plants growing wild (without man's help) on top of the Pryor Mountains in south central Montana.

Who adds fertilizers to this field? No one! Who waters this field? No one! Who removes the weeds?   No one!

A field of wildflowers in the Pryor Mountains

Very interesting. Where does this beautiful field of native wildflowers get its NPK (nitrogen, phosphorus and potassium), plus all the other nutrients needed to grow thousands and thousands of flowers? They must come from the soils. So where do the soils get all these nutrients? They must come from the old vegetation as it decays; all the dry and green plant materials (mulch) that lays on top of the soil where billions and billions of micro and macro organisms are consuming the rotting vegetation.

Hmmm, this sounds a lot like the process that we use for growing healthy foods. The answer is yes. It is God's natural plan, a natural form of sheet composting, which goes on and on, year after year without man's help, adding the needed nutrition to the top layers of the these very rocky native soils. Oh OK. So perhaps we should just copy this natural method of wild gardening and apply these same principles to our vegetable gardens. Simply layer homemade compost right smack dab on top of our native soils and add a thin layer of mulch several times throughout the year.

Yesir ree bob! That is exactly what we are doing with our soils. We then place seeds into the dark colored, soil-based compost and see what happens. Voila! Our home gardens methods are mimicking these natural native processes of this self-regenerating wildflower fields. No plowing, no money is needed.   Homegrown composting is the key to *Gardening for Life - where no money is required*. By the way, the blue flowers in this photo are lupine, a legume that fixes nitrogen in its root systems just as peas do in our garden beds, further improving the soils. The take home message here is simple: <u>we need to copy the natural process going on inside the soils within this field of wildflowers</u>.

Chapter 4. Garden Plans, Designs & Appliances

Examples of Small Garden Designs

Every garden should have a compost pile

African-Style Portable Food Garden

This idea was designed to help people that live in Shanty Towns in very small shacks, so they could wheel a garden in and out of their home.

I made this drawing on our trip back to the USA from South Africa while sitting on an airplane, March 30, 2008.

I was frustrated that our modern ideas would not help folks in South Africa people because of severe poverty - anything not nailed down is stolen. *Gardening for Life* actually started with this simple garden drawing.

We went into an extremely poor informal settlement (thousands of people crammed together in very small shacks) and measured how wide their doors were. The idea was to build a homemade wooden box mounted on wooden wheels with an A-frame rack suspended over the top of the small box garden. This A-frame held four buckets to raise bigger plants.

A person waters the top buckets then the nutrient- rich water drips onto the garden below. The idea was to move this portable garden inside their houses for safekeeping.

Anyway, here is the original drawing of that idea.

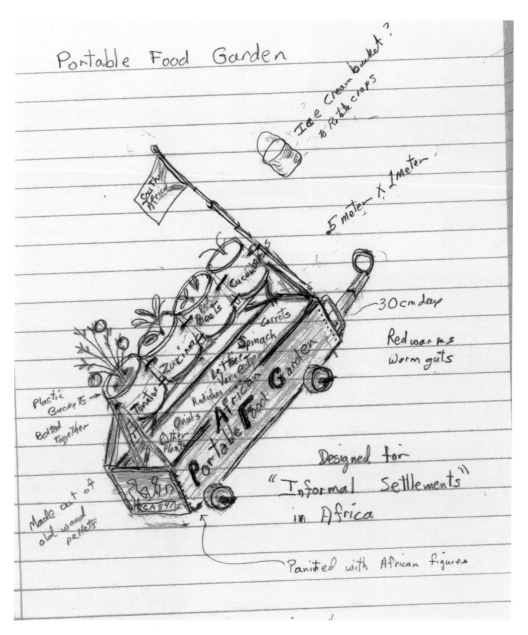

Step-by-Step Kitchen Garden Preparation

Step 1 Locate an area near your home that has full sunlight for most of the day (not under shade trees).

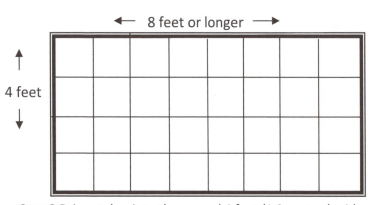

Step 2 Drive stakes into the ground 4 feet (1.3 meters) wide by how long you would like … 8 to 16 feet (2 to 5 meters) in length. Keep the width narrow so that your hands can reach the middle squares from the sides of the garden area.

Step 3 Dig down and remove all the roots and existing vegetation. Loosen the soils with a shovel or hoe. Line the outer edge of the box with logs, bricks or rocks. Fill the area with good top soil, mix in old livestock manure and compost.

Be sure to water before planting & let the garden set to settle the soil.

Mr Brite

Step 4 Add sticks or other material to make a grid with each square 1 foot by 1 foot (33 by 33 centimeters). Water the whole garden area and let sit for a few days to settle the soil. Never step on this garden soil, to prevent compaction. Next step… you are ready for planting.

Planting over 400 vegetables in a small Kitchen Garden

Note the different kinds of vegetables for health and healing

How to construct an Africa Style Garden at no cost

Step 1
Locate a full sun, handy spot

Step 2
Mark off a square 120 cm by 120 cm out of rocks

Step 3
Cover the bottom with cardboard if weedy

Step 4
Fill full of very good homemade compost at least 15 cm deep

Step 5
Divide into 16 small squares with rope, sticks or metal strips

Step 6
Plant each small square with 16, 9, 4 or 1 plants depending on the size of the mature plants

Permanent Garden Bed Design (with self-regenerating soils)

This is a cutaway view of one of our typical garden beds - most of our beds are four foot wide (1.3 meters) by as long as needed.   We have found that 8 to 12 feet (2 to 3 meters) in length is a handy size, which makes them easy to walk around and reach in to. We have one long box - 40 feet in length - designed to accommodate a drip irrigation system. Our multiple smaller boxes cause plumbing problems with drip systems - too many pipe fittings get in the way when connecting several garden beds together.

The best thing about this design is that the compost attracts earthworms, and they start to multiply. I have counted 60 earthworms per shovelful in two-year old garden beds. I have changed my garden care methods by directly placing leaves and kitchen waste into and onto these soils to feed the worms. I now keep watering my empty raised beds. In other words, I treat our garden beds as worm farms. I keep the worms well fed and they work for free and non-stop fertilizing the soil and drilling thousands of air holes that loosen the soils. I find gobs of earthworms just under the frozen soil during mid-winter. These beds are becoming self-regenerating, which makes them "no cost" garden soils, again, letting nature do the work for you!

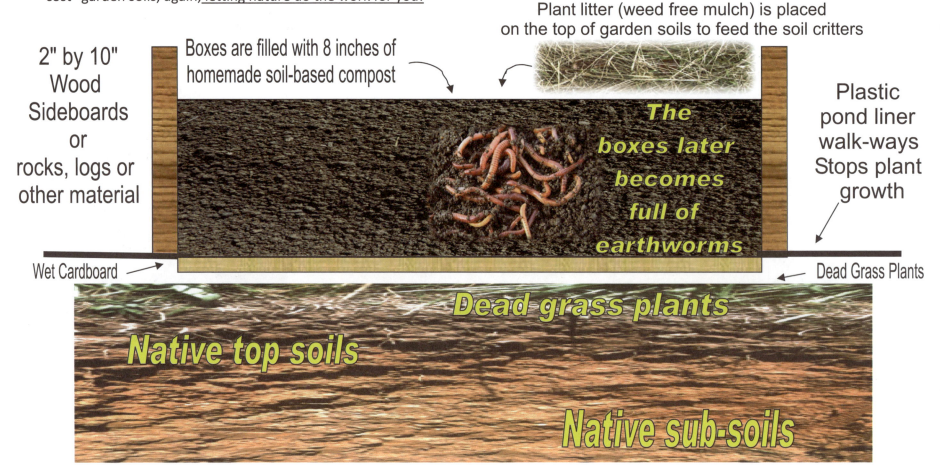

Bucket Gardening Test

Cabbage - Pepper - Potato -   Tomato - Squash - Sweet Potato - Corn

This was a test to see how well vegetable plants grow in 100% homemade compost. As you can see by the photo, all plants look healthy. They passed the compost test. However, the corn plants in the last bucket on the right made very small ears. Corn plants need much deeper soils, due to their extensive root system, than one 3-gallon bucket can provide.

The big advantage of bucket gardening is portability. If a storm comes along or is predicted, you can simply more the buckets to a safe location. Being a lazy farmer, I place them in a wagon and move several buckets around at once. In the fall of the year, you can place them in a heated area or bring them indoors, which greatly extends the growing time. We did this once with our pepper plants and the peppers turned red, which we had not experienced before due to Montana's short growing season (only 90 days).

Bucket Gardening --- A Portable Food Garden in a Bucket

Made with two buckets - fill one bucket with the soil and plants, then place top bucket into the bottom bucket which becomes the water reservoir.

Materials Needed and Instructions:

One 3-gallon plastic bucket (top bucket)
Drill ten ¼ inch holes in bottom to let water out
Drill one 3 inch hole in center of the bottom to place a cup in
Fill top bucket with a mix of 1/3 peat moss 1/3 mixed compost and 1/3 vermiculite
If material is not available, fill top bucket with mixed garden compost.

Another 3-gallon plastic bucket (bottom bucket) Used as water reservoir with one overflow hole drilled into side 2 inches from bottom of bucket.

One plastic or Styrofoam 8 oz. cup cut with slits in side to fit in center of bucket to wick water up into soil.

One 18 inch plastic PVC pipe 1" dia. Drill 1" hole in bottom of top bucket. Add pipe which is used to add water to water reservoir.

El-Cheap-O Way   Saves $$$$ A one bucket gardening method
Locate one 3 or 5 gal bucket. Drill a 1/4 inch hole on the lower side of the bucket about 1 inch from the bottom. Place 3 empty plastic water bottles with several holes or knife cuts in them to act as a water reservoir in the bottom of an empty bucket. Place a few rocks around the hole to keep it from plugging. Fill with good potting soil. Don't over water. Watch the side hole and do your best to never *let water run out this hole and you are good to go.*

***Success in buckets** filled with 100% good compost*
Cabbage, pepper, potato, tomato, squash, sweet potato, corn

How to Build a "Long Box" Garden - low-maintenance & no money spent

First off, why build a "Long Box" Garden?

After several years of growing food in small garden beds (4' by 4', 4' by 8', or 4 by 12'), I felt that building one long box 4' wide and 40' long would be much easier to take care of and become more productive, especially when considering the time it takes to water several small box garden beds. You see, for the last five years I hand-watered everything. One long box and one drip system could save me a lot of time and effort.

How to construct a 'Long-Box" Garden

Mark off an area where your new garden bed will be located. Stake out the edges and corners. The box should be only 4' or (1.2 m) wide by at least 40 feet (12 m) long, and 1 foot (30 cm) deep (about the height of a normal size shovel spade. Note the walk way is covered with used plastic pond liner to stop weed and grass encroachment.

Wet several sheets of cardboard and/or newspaper to cover the sod where you want to build the garden bed. Peel any tape off the cardboard and remove the staples. Haul in and spread homemade compost with equipment if possible. Afterwards, construct the box with 2 by 12" wood sides as in this photo.

You could wheelbarrow fill this raised bed garden box with your compost, but that is a lot of physical work. Either way there was no money spent on these materials for this garden box. All materials came from salvage.

Long box Garden

These are the before and after photos of the Long Box Garden. As you can see in the after photo, the squash plants took off, growing all over the place. Perhaps a box garden like this one is not the most efficient space to grow vine type winter squash plants. Next year, we will use the pothole gardening method described later in this chapter. However, we did enjoy lots of squash.

A "Wagon Hoe" Garden

This is a handy portable mini-greenhouse that is SELF-watering and SELF-temperature regulating.

Self-opening insulated cover that lets sunlight in.

The nifty GO WILD idea about this contraption is the covering. It's made with one layer of 1 mil plastic drop cloth, then strips of bubble wrap added as the next layer covered with another layer of 1 mil plastic sheeting.

Automatic drip system
Hanging on the back of this portable garden is one 3-gallon bucket with two drip hoses connected to a screen filter installed in the bottom of the bucket.

I wanted the wheels to be old wooden wagon wheels, but they were too expensive. I stumbled onto an old broken wheel chair, looked it over and made a few adaptations and bingo, I had a nice, very stable set of wheels that lets me roll this garden about, anywhere I would like to go.

That is the beauty of the "Wagon Hoe" garden. It's attractive enough to set outside our home and cute enough to cause a commotion when giving workshops, yet handy enough to grow food when old man frost comes about.

I installed a self-temperature- regulating lift arm, made with a black tube filled with beeswax. As the temperature goes up the beeswax simply expands, lifting the wire hoop with the bubble wrap covering. The lift is temperature adjustable. A nifty idea.

A portable mini-greenhouse

Instant Kitchen Gardens

This photo shows you how close the plant spacing can be with 12 beautifully diverse lettuce plants in a small pot measuring 10 inches in diameter by 5 inches deep.

Just place a few pots next to your kitchen door for handy eating. As anyone walks out the door, they can just snap off a few large leaves, check for bugs and if clean, pop the whole leaf directly into their mouths.

We call this, 'instant grazing' on healthy homegrown 'fast foods'. This is even faster than going to any fast food restaurant, plus no money is dropped off at

the drive-in-window. Works well for people on the run all the time.

The key to make this work is all in the soils. In order to grow beautiful food as in the photo, you must have healthy 'total-nutrient' soils. Plenty of NPK in the soils along with water holding ability such as perlite and peat moss make this a do-able fun project.

By harvesting the larger leaves, these pots become a 'pick and grow' garden. The hearts of each lettuce plant are not harvested which enables them to keep on producing.

We think that by locating a garden as close to your home as you can, you will use it even more. If you have to walk outside and around the back of a home, open a gate, duck under a bush, too much time has passed and in a busy fast-pasted world, you will tend the garden less and less. Make gardening as handy as possible.

Gardening in bucket is another great way to make food handy to harvest.

Pest Protection Ideas

Our first gardens in Africa started out fabulous. They were fast growing, healthy looking plants, and all the marked off squares were coming up solid wall to wall plants. The next morning we woke up and went out to check the garden and wow! Something had eaten all the newly planted beans, the lettuce was chewed down, and the tomatoes were missing several stems. What creepy crawly thing did this?

You see, we had not gardened in Africa before, so there, we were the new beginners. Adeline, the owner of this garden and a real African farm lady said, "Oh it's just the birds, Nazebazo" which means "no problem" in Kinyarwanda. The next thing we knew, she had placed wood sticks pegged into the soil with a mosquito net strung over the top of the plants. This stopped the birds. A sad beginning as some of the plants had to be replanted, but this is the way to learn.

"Mistakes are learning opportunities."

Mr Brite

As we learned more about how to keep pests from stealing our valuable food, we started experimenting with the following ideas:

A homemade, removable, plant protection cage, sitting on top of a small kitchen garden. Rwanda, Africa

A "Pumpkin House" built by an African farmer made from wet banana stems tied together, forming a tent over the plant.

Photo of the inside of the cage

A Salmon River Pumpkin well mulched with grass, all safe and sound from animals eating or stepping on this "Pumpkin House" cage.

Inexpensive Portable Greenhouse

Tote Box full of transplants sitting in the snow

This is a super simple idea to raise transplants in cold weather.

Start your tomatoes (or whatever you are growing) in 16 oz. cups filled with good clean compost and potting soils and place them in a clear plastic tote box.

This becomes a handy small movable greenhouse. If you are going to have a very hard freeze or are going away for a few days, just set this portable greenhouse in your home. Or, you can place a cover or lid on the tote, or even throw a old blanket over it when it becomes cold out. The high sides protect the plants from wind. You could even find ones with wheels on them for rolling in and out of your garage.

Say you wanted to cover it for a few days left outside; you could cover the whole box with bubble-wrap to protect it from a light frost. However, do not leave things covered too long or too tight, as moisture and high humidity is an ideal environment for aphid populations to explode and cause you other problems.

Also this is a great way to harden off transplants by moving them around to become acclimated to different environments before setting them out in your garden. Tender plants need different growing temperatures and some wind to become strong transplants.

To supply heat either use a heat mat made for growing trays, or try a string of Christmas tree lights as a heat source. Be sure to check things often with a thermometer.

Double Hoop-House

They say two are better than one. Below is a photo of a double mini hoop-house placed over a raised garden bed. The first layer is constructed with a floating row cover (a fine cloth-like material) and the second top layer is covered with three mil clear plastic sheeting. The stiff wire hoop is made from a concrete sidewalk reinforcement wire, which holds up this self-standing hoop. Each end is covered with a quilted blanket that adds additional warmth.

Half-inch plastic PVC pipe holds the floating row cover up. I just drilled three quarter inch holes in the sides of the wooden box and simply pushed the PVC pipe into the holes so that it forms a low arch over the transplanted seedlings below.

This garden bed has several transplanted lettuce varieties along with a 'shotgun seeding' of several additional mixed lettuce types, two kinds of spinach and a few radishes. This whole contraption, made with almost no money, was seeded on March 14 in Montana.

Montana's night temperatures do dip into the single digits. Just a week before this hoop was constructed it was only three degrees F early in the morning. I told my neighbor that I was just now planting our garden. He said, "Wayne, don't you know it's only March", meaning this is way too early to plant an outside garden.

I am not too worried, as each layer of protection adds about 10 degrees F. warmer temperatures. I seeded and transplanted only the very cold-hardy plants such as lettuce, spinach, kale and cabbage.

These new seedings will be OK, as we have spinach starting to grow in other hoop-houses outside despite the cold spring weather. These other spinach seeds were planted in our garden beds last fall.

My wife and I have a goal to eat a fresh salad from this garden in March. It will probably be baby green leaves. It is handy to place a transmitting thermometer inside this double hoop-house, which also records minimum and maximum temperatures. That way I can just sit in the warm house, glancing over to see what is cooking (too hot) or freezing (too cold) in this outdoor hoop-house. Ventilation is accomplished by opening each end on hot days.

The uncovered soil temperature on the day of planting was only 45 degrees F. The double hoop-house will warm these soils. Good news: as I planted each transplant, in each hole that I dug were several earthworms busily making free fertilizer. You see, last year I purposely fed and watered these worms, treating the garden beds just like a worm farm, and come early springtime these workers are providing us with healthy soils. This increased worm activity adds to our food security and makes these gardens beds self-regenerating.

Pot Hole Gardens

A pot hole garden is simply a hole dug somewhere handy that you can easily water. The pot hole gardens should be lined around the outside with materials that prevent weeds and from encroaching into the center where the plants are growing. You can use cardboard, plastic sacks, metal strips, rocks, old plastic pots or buckets with the bottoms cut out, rubber tires, wood boards, bamboo, or you can simply dig a dished-out hole and remove the unwanted plants. The outside added material act as a weed/grass barrier. You can use the boiling hot water trick to kill weeds around your pot hole gardens. Buckets with the bottoms removed that stick up above the soil surface also become windbreaks for small young plants planted in a pot hole garden.

Just fill in the holes with good top soil, mix in compost and plant seeds

Plants for Pot Holes
- Tomatoes
- Peppers
- Squash
- Pumpkins
- Cucumbers
- Vine crops
- Zucchini
- Plus others

Bottomless buckets act as wind breaks

Pot Hole Gardens in Africa

Pot Hole gardens also work well in hilly and rocky areas

Succession Block Planting

A common mistake gardeners make is over planting of a single crop. You end up with way more than you and your family can eat. The vegetables become old, too big, and bitter tasting. A good solution to this is block planting progressively as indicated in the drawing below.

We have over planted and ended up with a wheelbarrow full of wasted lettuce that became compost. Why grow compost this way - what a waste of time and resources. As we have learned to produce more and more food due to our great growing compost, I resorted to selling vegetables. At least I was not just turning all that effort into recycled vegetables.

Block planting grows a tremendous amount of food. Therefore, it is good to plan backwards. Say you want to be eating 2 bowls of lettuce a day. Plant one block that give you two weeks of fresh young leaves of mixed lettuce, and then in 2 weeks you plant in the next block. Our gardens produced 1.5 pounds of lettuce leaves per square foot. That is about 4 bowls full per square foot. Say we ate 32 bowls of lettuce in two weeks; we would need 8 square feet area every two weeks. That means we should plant blocks in two feet sections every other week, as our beds are 4 feet wide. We plant more than that as my wife loves to give lettuce to our friends and family.

Four lettuce varieties seeded in 1-foot blocks 4 feet wide

You can also cut lettuce and let it grow back several times during one planting.

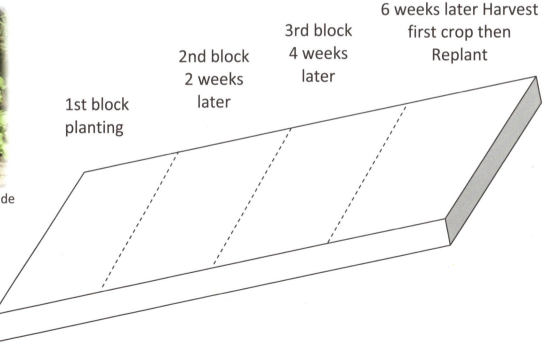

Ruth Stout's Gardening Without Work

I stumbled onto a gem of a mulching method for gardening. Ruth Stout, who died in 1980 at age 94, wrote articles and books documenting her methods. One in particular is titled "Gardening Without Work for the Aging, the Busy and the Indolent". She came up with a mulch gardening method. Her garden was simple: a fair sized garden patch completely covered with rotting hay and maintained that way year-round. She said that if a weed popped up, just put another arm full of hay on it.

This book is a treasure as she started gardening in 1930 and later started writing about her methods. She explained that she never plowed, spaded, cultivated, sprayed, dug, hoed, fertilized, weeded or built compost piles. In other words, **no work**. She just planted and picked. Plus, and a very BIG PLUS, she said she never had to water once, gardening in Connecticut. Hard to believe. Her first chapter in the book is entitled: "God Invented Mulch". Moreover, I thank God she started writing about all of this or her unique method may have been lost.

There is a YouTube video of her method that is worth watching. She said in the video that she feeds two people from her 45 by 50 foot garden, and not to make you jealous, had not been to a supermarket in 14 years. Does she have your attention? She sure has mine.

Think about what she was doing - feeding the soils with mulch on a year-round basis with all kinds of organic matter. In her book she wrote, "Please start with 8 inches of mulch". That thick of mulch will settle down flat and smother out most plants. She then just scraped away some of this thick mulch, planted a few seeds, patted the seeds down with her hand and ended up with a productive food garden that was much less work than most people would ever imagine. _Letting nature do the work for her._

Ruth once said, "At the age of 87 I grow vegetables for two people the year-round, doing all the work myself and freezing the surplus. I tend several flower beds, write a column every week, answer an awful lot of mail, do the housework and cooking - and never do any of these things after 11 o'clock in the morning!" In addition, she had a Roman style breakfast lying on the couch. WOW!

No Cost Gardens - a work less, grow more approach to raising low cost foods on a year-round basis even in cold climates.

Gardening for Life Intensive Small Gardening is a simple way that greatly increases your own food security. These methods use a deeper thinking, cubic inch way of building garden beds (10 to 12 inches deep) in which every cubic inch of your garden beds ends up growing some kind of food. Example: planting 36 seeds per square foot in a solid block spacing (carrots, onions, different kinds of leaf lettuce, radishes plus others) is the key to this unusually high production. The garden soils are enhanced with good, deep compost so that every cubic inch of root growing area becomes a soft fluffy growing environment providing highly fertile growing medium all the way down the root system. All grown organically.

For example, we can now grow over 1,000 carrots in a 4' by 8' area and enjoy plenty of easy-to-care-for delicious, large carrots. That means that 36 carrots are grown in one square foot with the individual carrot seed spacing at 2 inches apart. You simply use 3 fingers on each hand and poke 6 holes 6 times using both hands moving across one square foot and you have your correct plant spacing. Or for the lazy gardener, such as I am, I use the "shotgun seeding" method of sprinkling small seed from a jar with holes in the lid.

1,000 carrots were grown in this 4' by 8' box.

Large carrots were harvested in November

As we travel about the world teaching gardening, we have observed hundreds of different food garden designs and gardening methods. The old farm-style way of planting vegetables in rows still seems to prevail. Change comes hard, as row planting probably dates back thousands of years to an ox pulling a wooden plow.

There are three major reasons to move away from single rows gardens or still planting rows in garden boxes as we see in many community gardens.
1. You will have fewer weeds to control. When vegetables are seeded solid with no space between the plants, weeds are greatly suppressed, as no sunlight will be hitting bare soils when the plants grow big. The plants leaves become living mulch cover.
2. Valuable water is conserved. Once the plants start to mature all the soil surface is covered with shade, and the soils become cooler which reduces evaporation.
3. Production in greatly enhanced. Every square inch of enhanced garden soil space is growing food. Production is greatly increased.

Well-made healthy, diverse compost is the other key foundational element that makes cubic inch gardening work so well.

When people start viewing their gardens in 3 dimensional - cubic inches instead of a 2 dimensional surface - you start to understand why small garden can become extremely productive. It is very handy to keep the garden areas only 4 feet wide as this accommodates reaching any area of the garden bed without stepping on the soils. Compacted soils become unproductive growing environment.

The smaller 4-foot wide size garden beds make it possible to place wire hoops over each beds. You can build movable plastic greenhouses or shade houses over the beds and even use bubble wrap to help insulate fall and winter crops. You can use concrete reinforcement wire that comes in large rolls to keep animals out of your beds, or even attach small square plastic netting (with zip-lock ties) to these hoops to deter rabbits, cats and dogs.

Chapter 5. Planting Guidelines for Maximum Production

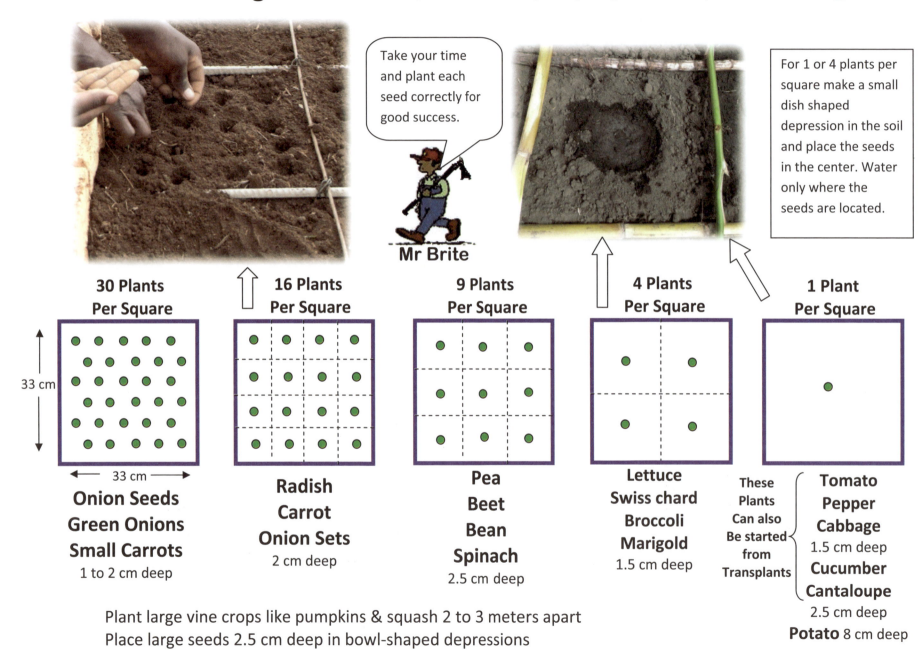

"Shotgun Seeding"

I once heard an intensive market farmer telling a story about the difference between conventional farming methods and his intensified form of agriculture. This farmer asked a conventional farmer, "How many rows of carrots do you plant in a 30-inch wide bed?" The conventional farmer answered, "Two". The intensive farmer explained, "In our beds, we plant 12-rows of carrots in a 30-inch wide bed. That is 8 more rows of carrots in the same space, which measures 2.5 inches between each row."

As a test for the last 3 years, I have planted in one square foot, on average, 30 germinated seeds of either green leaf plants or carrots. I have found that if I just hand-broadcast a whole gob of seeds in a 4-foot by 8-foot bed, it comes up about 2 inches between each germinated seed. In other words, it is wall-to-wall thick vegetable seedlings. I always tell myself, that I will just thin and eat. Somehow, deep inside of me, I struggle to kill even one plant. Therefore, I end up with a garden bed full of whatever I planted.

These beds were hand-seeded by simply holding the seeds in one hand and letting them fall slowly off the side of my hand as I moved my hand across each bed, a method referred to as "shotgun seeding". Meaning, I hope out of the mixed batch of seeds, something comes up. And sure enough, the beds look great.

There are several advantages to this kind of block seeding. One, I now have a very productive crop in one place, which makes it easy to harvest, water evenly and I can cover the whole bed with shade if need be. Two, it takes me only a few minutes to seed down one garden bed. And three, the plants come up so thick that one weed control effort a few weeks after germination is all I need to do, as once the vegetable crop really starts to grow and the plants touch each other, there is no room for weeds to get started and take over. It is simply too crowded. This is a form of biological control, plant against plant.

As I become more experienced, I am always trying to improve my methodology. I now use a small container with holes drilled into the lid. I then place a mixture of seeds in this container and just tap lightly on the sides of the container as the seeds fall onto the garden beds. This makes for a much more even distribution of seeds on the soil surface. In addition, this is a great way to re-use an empty spice container.

You can see some of the bigger seeds laying in the soil surface in the above photo. The light colored specks are spinach seeds, a variety called 'Prickle-seeded" spinach. I picked these seeds up from the Thomas Jefferson's garden store at Monticello, Virginia. At Monticello, they have restored our third US President's garden back to its original design and planted many of his original 1809 vegetables crops.

As you can see from the other photo to the left, I have quite a mixture of different kinds of seeds in this "shotgun seeding", another one of our tests - as we experiment with new ideas.

Fall Seeding for Spring Eating

This has now become a contest to see which salad we eat first come mid-March. Is it going to be the newly seeded plants started in February or the plants seeded outside last fall? Below are three photos of last year's fall seedlings taken in February. Remember this is Montana, and most commercial garden supply stores have not even geared up for the current gardening season. Today, as I look out the window it is snowing and we are still growing. This is a great time to test these kinds of GO WILD gardening ideas where we try to break a few rules. Like, do we following USDA's Zone 4 seeding guidelines? Or GO WILD learning new ways to feed the world.

Vegetables alive outside in Montana in February

A small spinach plant seeded in October   A larger spinach plant seeded in September   A mache (corn salad) plant seeded in October

All these February plants are growing outside (in reality, they are just staying alive with almost no growth during the very cold weather, as some nights are even below zero). They all have one layer of 3-mil clear plastic stapled to a wire hoop protecting them from cold weather, either in covered pit style garden or a small wire hoop-house. The plants do freeze at night, but do not die.

These two photos are views of our fall garden beds (seeded in early September) and how the hoop-house easily tips back for handy access. The lids are held in place by removable bungee cords to keep the wind from blowing them all the way to the neighbor's house, which is a quarter mile away.

We were able to harvest nice bowls of salads all the way to Christmas time. And we want to start eating again by mid-March which is harvesting salads for 10 months of the year. A great example of GO WILD! Gardening.

Backwards Planning Your Gardens

Before you put one single seed in soil, use the idea of 'gardening with the end in mind'. Try visualizing what you want the garden to look like the day you harvest it and note which day that will be. Often times we get big ideas during springtime dreaming looking at all the seed catalogs, and end up over planting, or causing too much work (weeding, watering, thinning, harvesting) later. Therefore, we should do our best to plan backwards.

An example of this is our desire to eat a salad from our garden daily from March to sometime in December, which is 10 months. Conventional gardens grow only 3 months of the year in Montana. If I plan backwards, this means that I need to start some lettuce seeds in the house by mid-February. As I write this page, I can turn around and see which seeds are growing under grow lights.

When these plants reach full transplant height, meaning the roots are sticking outside the bottom of the small starting trays, I will move some to the greenhouse. Others will be hardened off - I place a few starter plants in bigger containers, then place each lettuce plant in a large clear plastic tote boxes. I will start by setting the plastic box outside to toughen them up during the day and bring them in at night. The temperature fluctuation and wind will help the plants adjust. Once they are accustomed to this treatment, I will transplant them into raised beds that have hoop-houses over them.

Just for variety, I will also plant some mixed lettuce seeds outside in the hoop-houses by the end of February. This will extend the harvest timing; if I can stick to planning backwards, I need to be planting all spring, summer and fall in order to keep harvesting salad leaves when they hit their peak flavor.

A Healthy Lettuce Plant in a Plastic Cup

I do *'cut and let re-grow'* lettuce plants. We should just harvest the larger outside leaves, but often just cut off the whole plant at once.

When to Plant or Not to Plant?

That is a hard question to answer. There are certain kinds of cool weather plants such as lettuce, spinach, radish, arugula, Swiss chard, collards, strawberry spinach, broccoli, cauliflower, cabbage, beets, and carrots, plus several others, that go in the ground early. I start plants indoors as early as February and transplant them outside in March. Most end up in mini hoop-houses or some other kind of protected location such as rock wall gardens designed to hold heat at night.

However, I do say it is never too early nor too late to plant something. All these cool weather plants can again be planted in August and September for a fall garden. It burns me that most garden stores often pull all their seed packets for sale by mid-summer and a person cannot just stop by a hardware store and buy more vegetable seeds. You have to think and plan ahead for breaking the rules to Go Wild.... meaning to let nature do the work for you and copy how wild plants grow.

In addition, some fellows say you cannot grow this or that cold weather plant (like radishes or lettuce) mid-summer, as they will bolt in the heat, meaning the plant shoots a seed head and turns bitter tasting. I got around this concern by sorting through my seed packets looking for the slow to bolt plants. Then I tricked these heat-sensitive, cool weather plants by placing a shade-house over them.

Young lettuce plants growing well on a hot August day under a shade house

African Test Garden Before and After Photos

This was our very first "*Gardening for Life - No Money Required*" test. We called it the African Test Garden. This garden was constructed with no money. I just placed large rocks in a 4 by 4 food square as shown in the photo to the left.

I then filled the garden bed with 8 inches of our homemade compost made from local resources. I marked off 16 square feet as they do in Square Foot Gardening. I then planted several rare winter squash plants plus some other vegetables. The winter squash is an American heirloom plant that I am holding in the lower photo: Cucurbita maxima; cultivar: Lower Salmon River. We call it Salmon River Pumpkin.

This lower photo was taken 70 days after planting. Obviously, our compost by itself grew and grows a great garden.

70-days later

This first trial in the African Test Garden produced over 100 pounds of food. We consider this a break-through as some of the literature says you may have problems growing vegetables in straight compost.

We have tested growing in our homemade; soil based compost for 5 years without problems and experienced phenomenal results.

We actually have tremendous production in all our garden beds, but better yet is the fact that we came up with a method of growing healthy food with *no money required* that will work anywhere in the world. This no cost garden method is absolutely necessary in third world countries, as most people do not have the cash to spend on growing their own food. Their food comes from the wise use of local resources.

Plant in Multiple Garden Locations - don't put all your eggs in one basket

Gardens need backup planning. If one area around your yard shows slow production, other areas may just grow faster. Wind, the angle of the sun, tree and shrub location, building placement, fences, and rock walls all add to or subtract from a garden's microclimate. These areas can even change climates with different seasons of the year. Therefore, I plant the same crop in several different locations at the same time, just to see who outgrows whom.

For example, one year was really the best tomato-growing year that we have experienced. Daily temperature fluctuations are the dreaded downfall of gardening in northern climates zones around the world. This particular year we had warm nights well into late summer and the tomatoes ripened faster than they had in other years.

That year I planted 44 tomato plants of 13 different varieties in 5 different locations as a test to see which tomatoes grew the best. I placed some in buckets, others in the greenhouse, some outside in garden beds, some with bottomless bucket protection from the wind and others out in the open, still others in our raised garden beds. This was a great experiment to see which ones matured the fastest, grew the biggest crop and produced over the longest time. We had tomatoes galore.

The tomatoes growing in the buckets won the first-to-ripen contest. I asked a master gardener friend of ours, "What's the deal with the buckets producing the first ripe tomatoes?" She said that when their roots hit the bottom of the bucket, the plants ran out of nutrients and changed their growth pattern to producing fruit. That makes sense, as I know that the purpose of most everything on earth is to produce offspring for survival.

So, do not plant your favorite crop just in one location - instead plant in several different locations to see which one works the best for your micro-climate. This strategy also helps during a directional hailstorm; perhaps you will not lose your entire crop during single storm event.

Grandson with a tomato plant taller than he can reach

Chapter 6. Garden Care Made Easy

Precision Watering Ideas for Gardens

Do your best to only water the soils and not wet the leaves of young plants. A wet leaf increases the chance of plant diseases.

Homemade water jugs help you aim the water where you want it to go.

This lady in Shone, Ethiopia, Africa is a very good gardener as she knows how to place valuable water on each seed zone, which saves her much labor hauling hard-to-acquire water for her garden.

Plastic water bottles with holes poked in the lid make excellent sprinkle systems. They save water when you place the water only on the seed

Take your time and very gently place the water on the seed, don't wash the soils, which bury small seeds too deep or float them out of the ground.

Mr Brite

How to Use Compost

When compost has completed the active composting process, it appears dark colored, crumbly and soft, with a distinctive clear, earthy smell. You may want to sift out any solid sticks and large chunks of non-decomposed organic matter. This great-looking soil is ready to use. I love to feel and smell it as it flows through my fingers; must be that kid thing when playing in the soil. There are probably beneficial minerals in compost that your skin absorbs. As you can tell, we love compost.

Different uses for finished compost:

Top mulch
After gardens have been growing for a while, simply fill a bucket with good compost and keep it handy in your garden at all times. When the urge comes, spread a thin layer of new compost around the base of needy plants. Next, sprinkle some water on this added compost and the fresh nutrients will become suspended in the liquid and feed the plants and microorganism in your soils. It's hard to apply too much compost.

Add to potting mixes
I find it handy when transplanting small plants to larger pots, to fill the bottoms of the larger pots with good-looking damp compost. Make a hole in this compost and place the transplant in to the pot and top dress it with potting mix, which usually has been sterilized. Sterilized soil reduces the chance of mold forming on top of your soils.

You can sterilize your own compost in an oven at 180 degrees F for 30 minutes, in a microwave set on high for 2 to 5 minutes, or by letting the sun bake wet compost under a sealed clear plastic sheet for 4 weeks. Sterilization kills weed seeds and disease organisms that might be lingering in the soil. Do not over bake above 180 degrees F as this can damage the beneficial properties in the compost.

Side-dress existing plants
Each spring I dig a shallow trench around the base of some of our large perennial plants, such as rhubarb. I fill the trench with new compost and mix some into the existing soil. I then make a larger dish-shaped depression in the soil around the outer area of the plant, which helps hold the water in the root zone.

Make and use compost tea
1. Fill a cloth bag with your best compost. 2. Dunk the tied-closed bag into your watering can or bucket just like a tea bag. A stick tied into the knot of the cloth bag can be used to suspend the compost tea bag in the bucket of water. 3. Wait an hour or so as the water turns tea colored. 4. Remove the cloth bag and dump those contents into your garden beds just like mulch or back into the compost pile. 5. Water this newly made compost tea right away on the plants that may need some fresh nutrients. 6. **Watch the** plants perk up.

Feed the Soil and It Will Feed You!

Homemade compost in a bag

It is very important to feed the soils with good compost regularly. These organic fertilizers keep your plants strong and healthy. Actually, you are feeding all the organisms that live in the soils and they release minerals and locked up nutrients, which then become available to plants. Letting nature do the work for you.

If you don't **DO THIS** something called: "land idle-itis" will set in. The plants become stagnated; they lack good sturdy growth and become weak, making them more susceptible to diseases and insect damage.

Making and Using Compost Tea

It is very simple to make compost tea. A fast, easy way is to place 3 inches of good compost in the bottom of a 1-gallon glass jar. Fill the jar with water and let it sit for a day or so until the water turns to a rich brown tea color. Some people speed up this process by installing a fish tank bubbler that agitate the water and adds oxygen. The objective is to naturally make liquid fertilizer and mass-produce microorganisms that when added to your soils, promote healthy plant growth.

Another handy method is to place a porous cloth bag filled with good compost suspended by a stick tied into top portion of the bag in a jug of water, just like a tea bag. However, please label this jar "do not drink", as it looks like sun tea.

As plants grow, they use nutrients needing replacement. Compost tea is a process of nature that goes on continuously, every day, as moisture moves through a native plant community. Use homemade compost tea as a free (no cost to you except your time) organic fertilizer!

Compost suspended in a bag

Diluted compost tea ready to use

Plants benefit from added nutrients

The Best Totally Organic Weed Killer in the World

Boiling Hot Water! As we traveled the world teaching gardening, we did not yet have this trick. Darn, as when we were in Rwanda, Africa for 3 months, we watched our demonstration gardens closely and during the rainy season the grass on the outside of a garden bed started to creep through the rock boundaries into the vegetable garden. Weed control meant removing all the rocks, digging the weeds and grass out of the soil, even going after all the plant roots, and then putting all the rocks back - hard work!

This boiling hot water plant killing method would have saved the day in Africa. Below are photos documenting the treatment of hard to kill perennial white clover plant growing in one of our garden beds. White clover has a fast growing taproot. Interesting that within hours of the boiling water application, this clover was dead. Hot water has a lethal penetrating heat that can burn your skin and sure enough, it is a great organic plant control method. To treat unwanted plant with this hot water trick, move quickly and do not let the water cool down.

We would love to find more of these kinds of very low cost methods that make gardening much easier. "Gardening for Life - No Money Required" will continue searching the world for more ideas. If you have any of these low cost suggestions, we would love to hear from you. Just go online to our travel blog site at http://www.newwaystofeedtheworld.blogspot.com and leave us a message.

White clover plant growing inside our garden beds

Pouring hot water on this perennial plant

A dead plant within hours

I also treated a dandelion plant with the boiling hot water trick. Within a few minutes, the dandelion plant turned black and is no longer a problem. Dandelions have a deep, aggressive taproot system as noted in the photo below. They grow well in lawns and even after chopping them off with a hoe, they have the ability to grow back.

It is nice to simply pour boiling hot water on a dandelion and kill the taproot, which then becomes organic matter rotting back into the garden soils. This is a win - win situation by treating an unwanted plant and letting nature do the rest. Best yet, water is cheap, clean and organic. It just takes a little energy to heat the water.

Taproot system of newly emerging dandelions

Dandelion plant turned black within hours after hot water was applied

Biological Principles - That Make Your Garden Better

- Don't let your soils see daylight - mulch
- It's never too late nor too early to plant something
- If you don't feed your soils, they won't feed you
- When your plants talk -listen!
- Healthy soils grow healthy plants and when you eat healthy, you become healthy
- Idle-itis happens when plants and soils are not tended (sit idle for a long time), everything goes stagnant
- Plants need to be exercised which makes them stronger

- **<u>Don't let your soils see daylight - mulch</u>**

Ultraviolet sun light is a sterilizer. Nowadays, ultraviolet light is used in all kinds of situations to stop bacterial growth. However, when your soils become bare (naked of any kind of covering) direct sunlight has the ability to disrupt the DNA in living organisms and prevents them from multiplying. In other words, direct sunlight shining on your soils has the ability to kill your hard working 'Good-Guys' microorganisms. These microbes help build soil fertility.   Simple solution: mulch by chopping up some tree leaves or dry grass and use this organic matter to cover the surface of your soils.

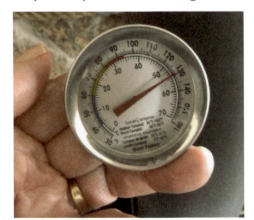

The sun's ultraviolet rays are also responsible for soil temperatures. Hot soils evaporate soil moisture quickly and there goes your precious water. Shading soils with mulch slows down this evaporation by blocking those direct sunlight rays.

I demonstrate this with a soil thermometer. On a hot sunny day, I stick the thermometer into bare soil and it reads 130 deg. F (54 deg C). Then I stick the thermometer into shaded soils and it reads 80 deg F (27 deg C). Soils that reach 130 deg F (54 deg C) kill (cooks) valuable soil building organisms.

- **It's never too late nor too early to plant something**

One of the bigger roadblocks that gardeners struggle with begins with the words, 'Yeah, but...I can't do that". People think that they cannot plant this or that plant, as they live in USDA Zone 4 or Zone 3, and there is not enough time left in the season. Or they think that it is too early or too late in the season to put a particular plant into the ground.

This is inside-the-box thinking. When you start thinking outside the box (non-conventional ways) things get exciting. I find that when I hear someone say, "You cannot do that!", my typical male ego response wants to prove them wrong.

I once went into a greenhouse and asked for some lettuce starter transplants. The owner said, "You don't need starter plants, you just seed lettuce directly into the soil". What she did not know was that we push our Zone 4 growing in Montana. Most folks grow for only 3 months, not extending healthy eating to 8 or 9 months of outside growing. But we cheat, using homemade mini hoop-houses to extend the season both early and late. We have a goal to eat salads from our outside garden starting in late March and continuing to munch on our good garden food until Christmas time.

The most effective way to eat this early is to start some of your plants (especially lettuce) indoors and set them out later. You see, this jump starts the plants and you get to enjoy fresh healthy salads when most people are still just thinking about planting their gardens. In addition, start some seeds outside under a hoop-house and see what happens. You will be pleasantly surprised.

Also, save some seeds from your hardiest fall/winter growing plants, as they are adapted to your local microclimate. These individual seeds possess a genetic advantage to survive in your specific growing environment.

- **If you don't feed your soils, they won't feed you**

Feeding your soils is similar to operating your piggy bank. When you keep removing resources and not putting something back to restore what you removed, your resources become depleted.

In our garden beds once a month we add new compost, worm castings and/or chopped up dry vegetation along with some chopped up green vegetable waste. This is food for the microbes and worms, which in turn give you more nutrients, released from the minerals locked up in your soil particles.

- **When your plants talk - Listen!**

Plants communicate with their appearance. When you walk into a room and a houseplant is drooped, it's saying, "I need a drink of water". Take a garden walk often and view what your plants and soils are telling you. Kneel down on their level with your nose on the ground and your rear in the air and observe. Just sit there for a while and talk to your plants, your breath gives off carbon dioxide, which plants love.

Look for signs of something unusual, like yellow or curled leaves, or bug holes. Pay attention to the soil surface: is it hot, dry, cracked, naked, or covered with moss? These are all indicators or symptoms that something is going wrong. It's important to remember that during your first time view, you are usually only seeing the symptoms of a problem. Don't treat symptoms, treat causes.

Symptoms are just indicators, telling you to dig deeper. Don't react immediately, instead sit there for awhile and ask yourself, why is this unusual looking thing occurring? Keep asking why, four or five times, or until you get to the real root cause of the problem.

A plant with wilting leaves.

The following is an example of finding the real cause or causes of a gardening problem.

On a garden walk, you may observe some pale looking cabbage plants. You ask the first why question... why are these leaves looking so sickly? You examine under the leaves and discover aphids. Before you treat the aphids, consider the aphids as still just a symptom of the whole problem.

Next, ask yourself, why are these aphids multiplying on the underside of these cabbage leaves? You discover that the humidity is just right to cause this no-male-needed insect population to explode. The third why question becomes, what is causing this high humidity? You may have found the real cause of this aphid problem. This part of the garden bed is receiving too much water. Over watering has set up a microclimate favorable to aphid populations. Short-term solution, spray a soapy water mixture on to the underside of the cabbage leaves. Long-term solution, dry up the environment where the aphids reside.

Aphids on Cabbage

- **Healthy soils grow healthy plants and when you eat healthy, you become healthy**

It just makes common sense that healthy soils are full of healthy nutrients, which move into the plants via their root systems. It is an amazing phenomenon that different plants have different amounts and kinds of vitamins, minerals, and other nutrients in them, yet they grow side by side in the same soils. Remember, we are what we eat!

- **Idle-itis   When plants and soils are not tended (sit idle for a long time), everything goes stagnant**

This principle shows up in all things. Just lie in a hospital bed for one month and see how strong you are then. As I have consulted on pasture management for 30 years, I can truthfully say; "Healthy grass lands need periodic disturbance and adequate rest to reach full potential". It is all about recycling, and earthworms do a great job of positive disturbance, as they chew through and move about, turning stagnating idle soils into nutrient rich soils. We can help these worms do their job by mulching and feeding kitchen waste buried into the soils, which gives them a thriving environment to multiply. If your soils lack earthworms, change something to encourage them as they work hard, which means less work for you to grow more.

As I have said before, a great gardener studies his plants, and soils on his garden walks, and learns how to adjust the microenvironment so that individual plants can produce a healthy, productive crop.

- **Plants need to be exercised which makes them stronger**

I once visited an astute gardener's greenhouse and he told me; "My tomato plants don't like to be touched", so every morning he walks through his greenhouse with his arms spread out wide and bumps all his tomato plants. He knows that by moving the plants back and forth with his hands and arms, it strengthens their stems. In addition, tomatoes are self-pollinating and the flowers need to be tapped (moved about) to release some pollen to self-pollinate the ovaries. So remember this principle to toughen your plants: movement is good, not bad.

Don't Weed – Cultivate (by shaving the soil surface with a thin blade)

Here is a time saving weeding trick that puts some fun back into gardening… get those weeds early! Old farmers used to say: "If you see it's time to weed, you are too late." Make it a goal not to let any weeds spread their roots deep and produce seeds anywhere near your garden.

"One weed goes to seed, makes 7 more years to weed."

Please cultivate early so you do not have to pull and dig large weeds later. If you have a large garden, you can lightly cultivate the soil surface holding a standup "soil shaver" tool as in the drawing below. Sweep the soil; by doing this your thumbs will be pointing upward, holding the tool and using a sweeping action instead of a chopping movement. It is much less work to go about sweeping just below the soil surface than chopping with a hoe. This method is described in detail in Eliot Coleman's book "The New Organic Grower".

Another important tip is that if you garden in boxes or raised beds and have planted in solid blocks instead of rows, the weeds have much less room to grow. You can make a very small cultivation tool out of a serrated kitchen steak knife. Just heat a spot on the blade and bend the knife blade in a 90-degree angle and you have a nifty small garden cultivation tool. See photo to right.

A Long-handle standup "Soil Shaver" Cultivator Tool 5 foot (150 cm) in length

A small kitchen knife with the blade bent at 90 degrees becomes a handy cultivating tool for small gardens

Why Mulch Your Soils?

Just Right!!

Too Hot!

Cool shaded soils = 22 deg C (72 deg F) = holds water better, adds soil nutrients, and slows weed germination

Hot bare soils - 55 deg C (130 deg F) = evaporate water fast, cook and kill valuable microorganisms, no added soil nutrients and weeds can germinate

Mr Brite — A smart farmer

"Don't let your soils see daylight"

What is Mulch?
Anything that you can lay around the base of your plants: dry grass without seeds is best, dry leaves, compost, dry crushed manure, even shredded newspaper.

Why Add Mulch?
Bare soils that have direct sunlight shining on them become very hot. Just go and touch them on a hot day. Mulched soils are covered as most soils in nature are covered. This does several things: greatly reduces soil temperature, reduces water evaporation, slows down weed germination, adds organic matter, which in turn is food for the soil and plants, and slows raindrop impact. It keeps the plants clean from dirt splashing due to raindrop impact.

Most important You are copying one of Mother Nature's natural soil building processes. Add Mulch, Add Mulch, Add Mulch

Celebrating Growth Through Your Mistakes

Celebrating your mistakes is one of the best ways to make needed change in your gardening methods to become more successful. The key word here is YOUR, that is YOUR mistakes. This is the way we all learn and make the greatest strides toward success. In this case, it's revealing to use cause and effect with biology. Your mistakes and your errors give you a chance to adjust what you plant, how you plant, where you plant and when you plant.

A good example of this is planting plants in Jamaica (a tropical Caribbean island), with very close plant spacing with the goal of producing the most food from very small area which became an error in this situation. This mistake deals with microclimates. High humidity tends to increase certain insect populations, and can compound the expansion of diseases in certain plants. Cabbages, lettuce, spinach and other leaf crops growing very close to each other hold surface water. These same plants spaced out do much better in highly humid climates. The cause - dense foliage, with the effect – increased high humidity, needed to be corrected. Individual vegetable plants with more space around them do much better in this climate.

Thickly planted lettuce - obviously the insects love this moist environment. Note the holes in the leaves.

The trade off is that it takes more ground to grow food but you don't have an increase in the bug population eating your vegetables or plant disease problems.

So please swallow your pride, like I am doing as I write this, and confess the mistake. This observation says to adjust your gardening ways and in this case plan to add more air space and sunshine around mature plants. Continue to always look for more mistakes. Isn't this how we all learn and improve things? Let's celebrate when we discover these kinds of mistakes that yield revelations.

This nice looking, speckled leaf lettuce plant is healthy and has no insect damage. Note there is plenty of room for air to circulate around this plant, letting the plant leaves dry out faster in this tropical climate. Increased sunlight around this plant dries out the space between the plants.

Garden Care Check List: Tips, tricks, and a list of common wasteful things you should stop doing.

Simple soil tests to grow healthier plants-- Pass these tests and you're on your way to healthy growing.
1. Soil structure, friable, good tilth - Touch, feel for crumb structure and find out how easy the soils breaks apart.
2. Compaction wire test - Push a stiff wire into your soil and see where it stops.
3. Workability, shovel test - Dig in your garden with a shovel, how many hard clods are there?
4. Soil organisms - Dig a 6 inch hole and count the living critters in this hole, 6 is good.
5. Earthworms - Dump one shovel full of soil on a flat area & count the # of earthworms, or build a worm cafe.
6. Plant residue - Take a handful of garden soil and spread it on a sheet to determine organic plant matter present.
7. Plant vigor - Observe health. When your plants talk, listen. Look for dark green leaf color, thick stems, and upright plants.
8. Depth of rooting - Think in cubic inches and ask what type of soil are your deeper roots growing in?
9. Water infiltration speed - Poor water on your soils and watch where it goes - into or run off the soil surface.
10. Water holding capacity - Days after you water, take a handful of your soil and see how wet it remains.
11. NPK levels - Use a kit or meter, or send in a soil sample for any nitrogen, phosphorus, or potassium deficiencies.
12. Soil Ph - Check your soil Ph level, or simply add lots of good diverse compost which helps correct Ph levels.

Garden appliances that assist with better garden care - the less work you have to do, the more fun you can enjoy. Most made from recycled or low cost tools.
1. Leave a bucket in the garden with a dipper and let the sun warm it to give your babies a quick warm drink
2. Fine-hole sprinklers (either drip, watering can, or wand with very small holes).
3. Bent knife to use as cultivator tool in small gardens.
4. Soil shaver hoe. Don't dig, instead shave the weeds.
5. Sharp shooter shovel to go after deep roots.
6. Leaf shedder to speed up decomposition.
7. Bag of dry leaves to cover small seeds & amend soil.
8. Bag of perlite or rice hulls to mix in to loosen hard soils.
9. Floating row cover for bug protection and to hold heat.
10. Shower Stall doors as wind break, winter garden.
11. Wire hoops to make a handy shade-house or green-house.
12. Long box with homemade drip system for easy watering.
13. Organic fertilizers (add phosphorus to tomatoes).
14. Compost tea (homemade miracle grow).
15. Worm juice (works like homemade miracle grow) diluted 20:1.
16. "Handy Andy" tote on wheels (tools remain in the garden at all times).
17. Carpet knee pads to keep those knees clean & comfortable.
18. Pond liner plastic or old carpet make good walk-ways.
19. 3 to 5 gallon buckets with bottoms cut out (wind break and bug shield).
20. Don't put all your eggs in one basket (plant same vegetables in different areas to prevent total plant failure).

Add Mulch for:
1. Birth control for weeds (prevents germination)
2. Makeup for the garden (improve looks)
3. Shock absorber for water drops
4. Acts like shade cloth to cool soils from the hot sun
5. Food for the beneficial small soil critters
6. Roof (shelter) for microbes
7. Water canteens for the plants & microbes (holds water)
8. Blanket for cold nights to hold temperature up

Important Rule: *"Don't let your soils see daylight"*- MULCH!

Do not do these things (a list of mistakes to avoid)
1. Letting plants and weeds go to seed in your garden beds (unless you are saving plant seeds).
2. Over watering and leaching soil nutrients away from your valuable plants.
3. Over-seeding without thinning, eat baby vegetables to thin and use Square Foot Gardening spacing grid.
4. Under-seeding - plants too far apart, creating wasted space.
5. Too big of a garden and not enough time to take really good care of everything.
6. Lack of ownership - write out an actual plan and identify the garden's main purpose.
7. Thinking only inside the conventional wisdom box (Go Wild and break some of the rules and ask yourself, why not?).
8. Lack of daily care - go on a daily garden walk and listen to your plants "talk".
9. Harvesting too late and/or no storage system in place.
10. Over planting or planting something you do not know how to use.
11. Applying too much nitrogen and end up growing lots of leaves with less fruit.
12. Do not over-till your soils; move away from using a rototiller, which just speeds oxidation of organic matter.
13. Shallow soils or different soil layers - Apply compost to clay and sandy soils, and mix these amended soils into the subsoil for gradual transition into the deeper subsoil (use a soil fork to break up this barrier). Earthworms will also help relieve parched water conditions in soil layers with differences in water permeability.

Lots of MONEY Required
$$$$$$$

No MONEY Spent

All made with local recycled resources

Conventional row gardens are not fun to take care of.
They require way too much work, tractor plows and rototillers, lots of weeding, take tons of water to irrigate, the sunlight hits bare soil killing needed microorganisms, and they require too much land.

A **Gardening for Life** 'Long Box" is fun to care for. No power equipment needed. It is made from used wood. All soils were homemade soil-based compost. It's much less work, requires less water, has less weeds and grows fast and will out-produce any row garden in a much smaller space.

How to Speed Up Seed Germination

Placing seeds outdoors in cold Montana soil that is below 50 degrees F. will add multiple days or even weeks to germination. If anyone is running short on time to get a garden started in early springtime, skip this early outdoor seeding. Instead, consider a warmer place to speed germination. I have waited three weeks to germinate spinach outside when I could cut that time by two thirds indoors.

Once my wife found an old glass jar in our cupboards full of dehydrated cherry tomatoes. They were cut in half, dried in a food dehydrator and stored in a rubber sealed glass jar for two years. She was about to throw them out.

I took a couple of these Sweet 100 cherry tomatoes, soaked them in a jar of water for a few hours, then washed the seeds in a wire strainer. I ended up with several seeds to experiment with answering the question, are these seeds still good?

Cherry tomato seeds germinated on a paper towel in only three days. When transplanting, cut the paper towel with seeds in place to prevent root damage.

I placed several tomato seeds on a wet paper towel, rolled up the towel up, and put the whole thing in a sealed plastic bag.

I then placed this bag above our wood fire-heating stove. Three days later, I peeked into the still-wet towel and sure enough, more than half of them had germinated. The photo above shows the seeds with their white colored roots growing into the fibers of the wet paper towel.

It normally takes me over one week to germinate tomato seeds, so this warming technique cut that time in half.

Next I transplanted these seeds with their root still stuck into the paper towel. I just cut the paper towel with a pair of scissors and placed the whole seed, roots and paper into a transplant pot (16 oz plastic cup) full of potting mix.

You can tell if your tomato plants are healthy or not by their short stout growth and dark color green leaves like this one in the photo to the left. Tomatoes are a tropical plant and I will wait for warm weather usually protecting them with a jacket in a "wall -of-water" to help hold the temperature up even early June, as we live at 4,440 feet in elevation in USDA zone 4.

A healthy tomato plant, 4 weeks old growing in a 16 oz plastic cup with holes drilled into the bottom for water drainage

No Water - No Problem   Not Home - No Worries

Here are some simple solutions to watering problems. One is deep mulch (lots of organic plant matter piled high upon the soil surface). I once left 6 inches of dry leaves on a garden bed through the winter season. The following droughty spring the naked ground (no covering) was dry as a bone and the deep mulched beds held so much water that is was soggy, heavy wet soil. These areas were located side by side.

On a trip to Africa, no one was at our home to tend the small plants for several days, so I just piled on the mulch as deep as I could around the outside edges of the small newly planted vegetables and everything stayed just fine.

In many third world countries they have a wet season and a dry season. Water is a huge concern; some people spend hours each day hauling precious water considerable distances to their homes. There is no water for a garden. But here is another excellent solution. Instead of dumping kitchen rinse water or clothes wash water on the ground, just dump this valuable water on a small kitchen garden.

Soap in the water will not hurt plant growth. Certain soap contains phosphate that is good for plant growth, plus soapy water is often used for insect control on green plants. On a three month stay in Rwanda, Africa, we watered a garden with soapy water and it produced abundant vegetable crops.

Deep mulching and recycling wash water will get rid of these sayings, "I cannot have a food garden, as we have no water". Or, "We are gone from home too often to have a vegetable garden". These nifty ideas can break those 'Yeah Buts' that hold people in ruts.

Pour waste water on a small garden

86

Helpful Planting and Watering Ideas

Locate a used large clear plastic water bottle to use as a watering sprinkle can. Punch or drill 3 to 5 small holes in the lid, fill the bottle full of water, replace the lid, hold bottle upright over the plants and squeeze. Aim this fine stream of water toward the root systems of the plants; do not add water to the whole area. This saves valuable water, plus slows down the germination of weed seeds.

When watering newly planted seeds, be very careful not to use too much water, as this tends to float the seeds up to the soil surface.

African people demonstrated a helpful trick to accomplish this, by sticking a few green leaves with stems in the end of a watering can.

This slowed down the flow and spread the water more evenly around the planted area only.

The above photo illustrates how to carefully plant small plants such as carrot seeds in squares with close spacing.

Mr. Brite says the following:

Some Gardening Do's and Don'ts
Mistakes to avoid when raising your own food:

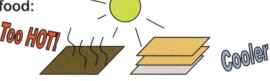

1. Don't let your soils see daylight. (said over and over, so just do it)
DO cover bare soils with mulch as this cools the soils on a hot day, holds moisture longer, adds organic matter, slows weed germination and keeps the soils warm at night like an insulated blanket.

2. Don't feed the poor looking plant first. (avoid just treating a plant with sprays first)
DO check and treat the soil first, <u>because if you don't feed the soil, the soils won't feed the plants that feed you</u>. Use black gold - add good compost.

3. Don't over-water the garden as it just leaches all the soil nutrients away.
 DO keep soil moist.

4. Don't over-till or dig up the soils too often as this just speeds up the decomposition of organic matter and sends carbon up into the air.
DO add good compost (black gold) back into the soil where the carbon is needed.

5. Don't burn old vegetation, as it sends valuable carbon energy the wrong way... up into the air where it changes the weather.
Do make compost out of all organic matter so that the carbon goes back into the soils and can grow more food for you and your neighbors.

6. Don't let your plants become too old and over-mature, as they don't taste so great.
Do eat them while they are young and tender. Re-plant as often as once per week.

7. Don't let any of your blocks in your garden go empty. Mulch if empty or
Do re-plant weekly by just cleaning the harvested plant out of the space, add a scoop of compost and plant a new crop.

8. Don't step on your garden soils, as this just compacts the soil and makes it hard for plant's root system to grow downward.
Do build your garden blocks small, so you can reach the entire garden from small paths and not walk in your garden soil.

9. Don't let your weeds grow big and or go to seed anywhere near your garden
DO cultivate often, don't chop-hoe the weeds. It's better to shave the soil with a sharp tool or pull the very small unwanted plants out of the soft soils when they are young.

Chapter 7. Year-round Growing Ideas - Even in Cold Weather

Freezing Weather Gardening

I checked the air and soil temperatures in our outside garden beds during a dreadfully cold Montana morning in late March and discovered some interesting observations. Spinach, kale, garlic, mache, and strawberry spinach are extreme cold weather plants as they actually freeze, but do not die in freezing weather even as low as 6 degrees F. which is much below "just a frosty morning".

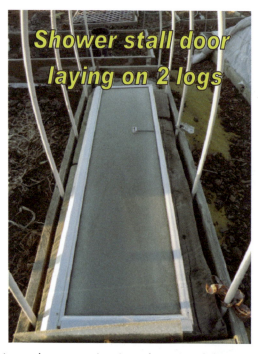

Upon close examination, the spinach leaves were stiff, hard and frozen, but as soon as the sun came out, they thawed and were quite edible (that is spinach in my mouth in the above photo on the right). The plants in the center photo were seeded last fall and had just starting to grow in the warmer March weather having made it through the entire winter as very small plants under protection from one glass shower stall door.

Soil temperatures ranged from 23 to 31 degrees F. The composted soils mulched with fall leaves were 10 degrees warmer than the uncovered native clay soils. The take home message here is that cold weather plants can adjust the liquid inside their leaves with plant antifreeze. The literature says these plants convert starchy liquid inside the plant cells to sugar, which does not rupture like water would, inside the plants. In addition, this is why carrots are much sweeter in the wintertime than during summer growing conditions.

Cold weather hoop-houses let us eat 9 months from our garden beds

This is a way to best Mother Nature, by growing with a protective plastic layer, which keeps soils warm at night. In our test garden we can now grow and eat cold weather plants (lettuce, spinach, kale, arugula plus others) for 9 extended months here in Montana's very cold climate. There is one big advantage of growing food early spring and late fall in a cold climate - **NO BUGS!** It is too cold for them.

These are wire hoops with attached 3 mil plastic drop cloth attached with paper stables at the bottom edge of each side of this mini hoop-house.

The hoop-houses are placed over a raised garden bed and secured with a bungee cord attached to the wooden sides of each bed and stretched over the top of the hoop. Simply tip them back for access.

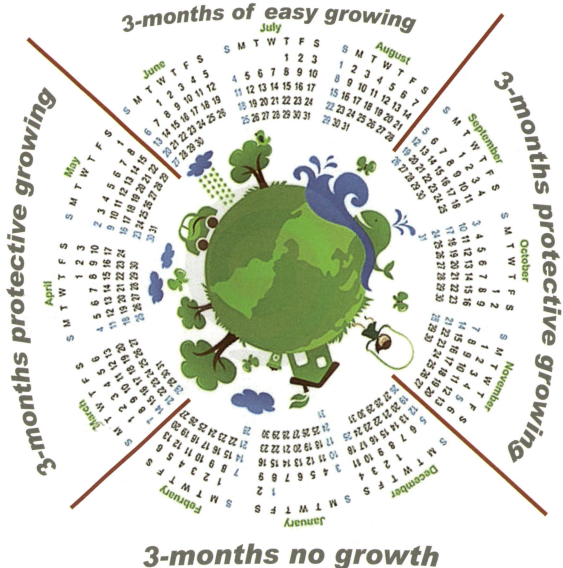

3-months of easy growing

3-months protective growing

3-months protective growing

3-months no growth

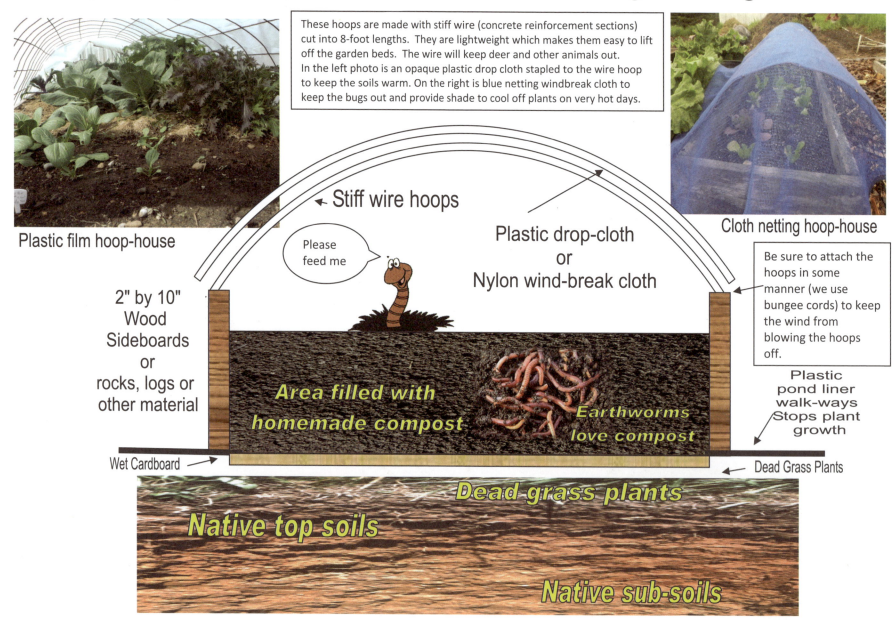

Wintertime Carrots dug mid-February in Montana

These are a variety of yellow carrots, which are less sweet than the orange varieties.

Note the dark colored soils (humus present). These soils are insulated with a 4 - 6 inch layer of leaves with a clear plastic cover over the wintertime carrot bed.

These soils were not frozen as of mid-February, probably due to climatic changes.

Worms and microbes are still very active in these unfrozen soils. This is a nifty way to regenerate soils, a form of re-composting once used composted soils.

Steps to growing wintertime carrots
#1. 1,000 carrots seeds were hand broadcast thick in a raised bed box garden 4' wide and 8' long.

#2. Come fall - harvest carrots as needed. When the weather turns very cold, cover the carrots with a layer of 4 to 5 inches (dry organic matter such as fall leaves).

#3. Cover the leaves with a heavy clear plastic. We used a old shower curtain. The thickness of a shower curtain helps keep the plastic from ripping during cold windy wintertime weather.

#4. Place weights on the shower curtain to keep the wind from blowing it away. This covering also holds the dry leaves in place.

#5. Simply flip back the cover and dig some tasty, crisp carrots all winter long. Sometimes the soils will freeze, but if dug soon after a very cold spell, they may still be usable. If left too long in frozen soils, the carrots become soft.

There is a big plus to letting the carrots grow on into the wintertime. Inside the carrots themselves, the liquids start to turn to sugar (plant antifreeze) and the carrots become much sweeter.

Growing Microgreens Year-round

Red Cabbage   Sunflower   Radish   Mix Greens

MICROGREENS. These are vegetable seeds (sown very thick) growing in trays or pots sitting in your sunny window all winter long. You can eat them in just 8 days. Sunflower microgreens, our best tested, can be germinated in the dark in a water bath for 12 hours, covered for one or two more days in a bowl with holes in the bottom, and then planted on the 3rd day in 1 to 1.5 inches of potting soil or compost. In just a few days, you will be eating a very tasty treat that is high in protein. One double handful of sunflower microgreens weighing 3.5 oz could have 15 to 22 grams of protein in it. You can top-dress soups, make whole salads and add to microgreens to sandwiches, or simply eat raw these yummy life healing foods.

Below are Black Oil Sunflower seeds ready to eat in 8 days.

Day 1 ... Seeds   **Day 3 ... Planting**   **Day 6 ... Growing**   **Day 8 ... Eating**

Steps to Grow Your Own Microgreens

Day 1 Seed heavy in tray of compost   Day 1-3 keep covered - inverted tray   Day 4 uncover   Day 4 - 7 grow in sunlight window

Keep moist.   Use a spray bottle at least once per day and more often when necessary. Do not over-water as this causes plants to mold. To help preventing over-watering use two trays, one with holes in the bottom, and another without holes as a water reservoir. Watch daily & adjust.

Day 8 Harvest by cutting with scissors, refrigerate in open plastic bags

Day 8 - 14 Serve, Eat and Enjoy   Replant new tray

Wintertime Growing Vegetables with Two Shower-Stall Door

Two bathroom shower-stall doors leaned together make one of our best all winter gardens. This whole effort started several years back as I kept walking by the two doors leaning together and noted that the weeds stayed green all winter long. The following year I planted some spinach and lettuce inside this contraption and sure enough, they too stayed green all winter long.

The two shower-stall doors make the handiest outdoor winter protection in our entire garden, as you just lift the door open for picking, watering and checking. In addition, the wind does not threaten to blow the whole thing away as can happen with some of our plastic covered hoop-houses. We live where the wind often howls at 60 miles per hour.

I check the air and soil temperature in this protected environment and to my surprise the air temperature stays about the same as the outside air temperature. The soil temperature is what stays warmer. Our winter plants freeze stiff as a board but do not die. The real cold weather champions are beets, carrots, collards, kale, arugula, spinach, and mache.

I leave one side of this appliance open all summer and the other side becomes a windbreak.

We love it, as it is such a simple, no cost way to grow almost all year long. When the below zero weather hits, the plants just stop their growth and wait for some warmer weather.

Chapter 8. GO WILD! Ideas - where we get to break the rules

What Does GO WILD Mean?

The words "GO WILD" refer to a stretching or breaking of common gardening rules. That is, GO WILD! For example, many gardening books tell us not to plant until the soils have warmed up to above 50 degrees F. Yet, we have seeded beds in soils that have tested only 28 degrees F. Certain cool season vegetable plants germinated very well in these cold soils and produced an early crop. After all, in nature there are billions of seeds scattered about the soil surface waiting all winter to germinate come warmer temperatures.

I have a theory that when you see any green plant beginning to grow, such as lawn grass poking green blades of grass up on a south facing slope, it's time to plant something. Let nature tell you when to put seeds in the soils or when to transplant. Better yet, plant spring crops in the fall and just as nature does, come early spring you have a garden bed full of fantastic early spinach and mixed lettuce. Certain cold weather vegetables act as winter annuals, they lay green all winter long and start fast growth early in the spring, just like dandelions.

"Shotgun Seeding" in blocks that make gardening fast, easy, and simple, all plants growing in 100% homemade compost
Shotgun Seeding is explained in Chapter 5

GO WILD also means studying a wildflower patch and copying the biological principles that are active on a year-round basis, and then applying these same processes to our garden beds... letting nature do your work for you. For example, tree leaves on top of your garden soils in the fall. Throw some compost on top of these leaves, otherwise they may blow away, but do not work them into the soils. You are copying nature as you watch tree leaves float to the ground and dust and dirt cover them.

GO WILD happened to me during one nice springtime day. After having added 4 inches of leaves to all our garden beds in the fall, the following spring, I was going to use a small electric tiller to mix up the soils with these old leaves that were just lying on top of these beds. Upon closer examination, I noticed that gobs of earthworms were active even when the cold springtime air temperatures were as low as 3 degrees F. Tilling these soils would just kill thousands of these earthworms.

I had to stop myself from working these soils, just as everyone else typically does in the springtime preparing the soil for planting. You see, I want to garden nature's way GO WILD, letting nature do the work, work less to grow more, and enjoy my time planting and picking more than tilling, weeding and watering. Therefore, I am trying hard to un-train myself from my old ways.

In review, here are ideas that we consider GO WILD:

Spinach plants that stay alive all winter long

Solid block fall planting that maximized production and stopped weeds from growing. Growing in a mini hoop-house all the way to Christmas

The Lazarus Tomato trick

How to jump-start tomatoes and speed the time to harvest

How to do it

#1 Break off sucker stem with leaves (a new branch growing off the main stem)

#2 Poke the cut off stem in a bucket of wet potting soil

#3 Keep the soil damp and watch it grow

#4 Drill a small 1/4 inch, water drain hole in the side of bucket, one inch from bottom to drain water

July 29

SAME PLANT BEFORE & AFTER PHOTO
48 days later tomatoes + new flowers with 24 inches of new growth

Sept 16

Deep Mulching a Green Solution

Holding adequate water and fertility in the soil especially in droughty times can become challenging. However, while in Mozambique, Africa, Antonio Aljofre director of the CNFA Farmer to Farmer programs, came up with the words, "Green Solution". He was referring to a large pile of cut vegetation in the right side of the below photo. Deep mulching with dry grass piled on top of the soil surface will provide shade and protection from the drying effects of sunlight energy.

The bare soil on the left side in this photo, that the African school kids are observing, will dry out very quickly and turn into dead soils requiring a lot of water and fertilizer to grow a productive crop.

The deep dry vegetation on the other side is slowly decomposing, feeding the valuable soil microorganisms.

To encourage the kids even more, we told them that in their eyes they were probably seeing dead grass that needs to be burned. In our eyes we could see money. That cut dry grass could be turned into valuable products such as soil carbon in the form of humus which becomes part of the soil. This dry grass could either be left in place, directly planted into it by parting the dry vegetation or piled into large heaps making "Black Gold" compost.

We also discussed the fact that the world's population is dramatically increasing. For the fun of it, I told the students that the cell phones that they were carrying were uneatable. Not saying cell phones were bad, but that we better know where our food comes from. Just knowing how to grow food, fast, productive, and healthy and not burning this valuable dry grass could someday save someone's life by having more food at less cost (saving the cost of commercial fertilizers). We should all stop burning which is sending carbon the wrong direction upward, polluting the air we breathe; instead we must send this life giving carbon downward, stored in our soils to raise healthy organic vegetables.

Here is a bumper sticker that we should all start thinking more about. The future of the world is in our hands.

Save a Person's Life.   Stop Burning!   Make Compost

How to Kick-Start Your Old Garden Beds

Here is another *GO WILD!* Gardening idea that we call: "Blanket Mulch Composting".

We consider this nifty new idea a gardening breakthrough. We have struggled to come up with ways to re-generate used compost.

You simply take about 2 inches of chopped up, mixed and/or shredded dry plant material and green plant material along with some nice rich soil, mixed all together, and apply right on top of your garden beds. It is putting your beds to rest properly. This idea will help you to follow the biological rule: **"DON'T LET YOUR SOILS SEE DAYLIGHT".**

This mixed up mulch feels and smells a good deal like compost. You can scratch it into your soils so the wind will not blow it away. Also add a bit of water on this newly applied mulch and let it sheet compost. This idea copies nature in the way trees drops all their leaves in the fall.

Our garden beds should be in much better from next spring with this nice looking, added organic matter lying on top of our used garden beds. You see, we grow in almost 100% soil-based compost with great results, and we know that some of the nutrients have been used and converted into nutritious vegetables we have harvested. In addition, this new idea will help to accomplish this biological rule**: "IF YOU DON'T FEED THE SOILS, THEY WON'T FEED YOU".**

Stationary leaf and grass shredder

The first year we did this was a success. I transplanted several cold season vegetable plants into one of our garden beds. Each hole I dug was full of earthworms. I do not have to amend these soils as the earthworms are doing that job for me.

This works like a weed eater with nylon whip blades.
If you do not have one of these machines, just chop up your dry and wet material on a block of wood with a ax or machete.

Hand chopped works just fine.

Build a "WORM CAFE" and your workers will come to you

How to build a "Worm Cafe"

Find a plastic jar, drill 3/8" (1cm) holes in the sides and bottom about 1" 2.5 cm apart. Leave the lid off. We learned that keeping the lid on caused too much humidity in the jar and it grew mushrooms.

Add some worm chow to your soils. Worm chow can be any of the following: vegetable waste, a little corn meal, carrot tops, watermelon rinds, anything that will rot and the worms love to eat. Mix the worm chow into the soil. Fill the jar 1.4 full of the soil/worm chow mix.

This jar of food becomes a *"WORM CAFE"*.

Bury this jar in your soil where you think some worms may reside. Add water once in a while to give the worms something to drink.

After 3 weeks or so, dig the jar up, dump it out and see how many worms you find!

Amazing! As the worms crawl though the small holes into this plastic jar, you have them trapped. Dump it out and count your worms.

Next, put them to work in your garden soils or add them to your compost and they will help you to begin building self-regenerating **"black gold soils"**.

We call this **better than organic** because these worms **labor** for you free and add some of the **world's best** fertilizer to your soils.

This is a great way to see if your soils have earthworms

Create Worm Farms Right in Your Garden Beds

This is one of our most successful GO WILD successes in our attempts to make gardening simple, easy and much less work. The photo below is one of our 4' by 8' raised beds come springtime after we fall mulched it with tree leaves. You can still see the leaves on the soil surface.

Note the earthworms are near the surface, probably feeding on the rotting (composting) leaves. It freezes every night, yet the earthworms are very active. This photo was taken early in April.

The GO WILD deal here is ... we are not going to apply any kind of organic fertilizers or other soil amendments to this bed. We want to see how well plants grow in a GO WILD fashion just letting nature do its thing. We will apply organic nitrogen, phosphorous and potassium to other beds and conduct a comparisons test.

This is how we are letting our soils re-generate themselves. Gone wild!

Chapter 9. Teaching and Sharing with Others
How to grow their own health

Teach Them Young

In today's fast-paced world, our young folks are missing out on how the older, slower-paced life feels. In my childhood back in the 1940s, I can remember that every home seemed to have a backyard garden. I still carry fond memories of my grandmother's garden, eating fresh peas shelled right into my mouth. These very early influential years had a large impact on my life.

I now feel it is up to us, that is all of us who can, to please introduce any young person that will pay the slightest bit of attention to an experience of seeing where food comes from. I have even observed, to my horror, our own grandkids thinking a freshly peeled potato was an onion, or that a whole chicken sitting on the kitchen table was not really a chicken. After all, it **did not** look like chicken nuggets.

It going to be a tough world out there, if the trend is true, and fast, convenient food continues to dominate our way of living. We will become more and more dependent on corporations to feed us, and lose the life saving knowledge and skills about raising and eating real homegrown food.

My wife and I wish that anyone reading these words right now would go out of their way in the next few days and teach a few youths something about growing homegrown food.

We encourage you to:

- Plant a kid's garden or better yet, let them plant one
- Plow up the front of a schoolyard and build get healthy demonstration gardens with flowers and vegetables
- Give a talk at your local school, church, community function
- Drag a wagon full of good looking vegetables in a parade
- Set up a fun interactive display for kids at a Farmers Market

   DO WHAT YOU CAN!

Teach and Tell Others About the Health Benefits of Growing Your Own Food

Don't overlook any chance to teach young people about how to raise their own food. Give them a square to be responsible for and place their name on it. Ownership motivates.

You are welcome to use this photographic garden manual as a tool to explain how easy it can be to grow your own food. Pictures communicate in any language. This is Connie, my wife, teaching the finer skills of gardening in Ethiopia.

Planting seeds of knowledge for those in need.

Teach a lesson and then let them teach others the same lesson

Young People Are the Future

It is up to us to take any chance we have to teach and demonstrate to the adults of tomorrow where healthy food comes from. Youth of today know about stores and fast, convenient foods. They are missing the importance of where real food (healthy food) comes from. Real food comes right out of the earth.

Feed them right out of the garden so that they can experience going from a seed to a plate to "Hookem Young".

Fun Ideas for Teaching Kids About Gardening   By TJ Wierenga

Nature Journal
Make or buy a simple notebook and show your kids how to write and/or draw an "experiment" of growing a plant from seed or seedling. Document, illustrate and date each stage: planting, measuring growth each week, first flowers, fruit/harvest. Use your Journal to get outside and get in to nature!

Taste Discrimination
Grow and taste-test a variety of cultivars – does your family prefer yellow, purple, red or standard orange carrots? Which lettuce is their favorite? What about "Yellow Pear" or "Sweet 100" cherry tomatoes? Teach kids to think about what food tastes like, to savor the differences and have favorites.

The Nose Knows
Make a game out of blindfolding family members (even mom and dad!) and trying to guess the names of a variety of herbs grown at home. Is it rosemary? Basil? Parsley? Chives? Mint?

Let Them Grow It
Give your child a small garden spot of his or her own, no larger than 2'x 4' for their first garden, or for young children. They need to be able to reach into it easily, and there shouldn't be so much garden that it is overwhelming for a first-timer. Draw out a diagram for young children, or write it out for older ones, showing the layers of newspaper, compost, mulch... and let them help you build it that way. Help them plant some seeds and seedlings both of "kid sized" veggies and fruit. Seeds such as carrots, mild radishes, mild loose-leaf lettuce, or seedlings such as a cherry tomato vine, or a couple of strawberry plants for some fun fruit. Plant and let a pumpkin vine sprawl somewhere!

Feed the Worms
Many children really enjoy the "bug angle" of gardening. Earthworms are fun! Let your kids help you make a Worm Café, from figuring out what the worms might want to eat from your family's trimmings, crusts, and excess... to burying the café, and coming back later to count the helpful visitors. Kids who love fishing will particularly enjoy realizing they can "grow" enough wiggly worms that a few can be harvested occasionally to draw in a fresh, flopping river trout!

Selling Produce is FUN!

Entrepreneurs
Help your child market extra veggies to extended family, friends or neighbors. Consider participating in a local farmer's market, or even a roadside produce stand (along the lines of a lemonade stand). The benefits of this are in so many areas, from social skill development, to recognizing the good qualities of their own produce, to money management. Just getting them interested in interacting with the world, and growing enough to sell, is a huge plus!

Be Creative
There is much joy and pleasure to be found in gardening. Use the "no work" methods outlined in this book to make your and your children's experience both fun and worthwhile!

Feed the Compost
Compost piles can vary widely in style and size, from round wire enclosures to unenclosed piles, compost balls that roll around the yard, tumblers that keep ingredients stirred up, or simple plastic containers set in the garden. Involve your kids with setting up a composter that fits your yard and budget, then charge the kids with keeping it "fed". A small plastic container or bucket sitting on the countertop will quickly fill with produce scraps and ends, bread crusts, peelings and the like. It makes a great, and rather fun, chore.

Garden Party
Have a summer or early fall birthday child? Consider throwing a party near the garden, and have your child help you decide what to serve. Fresh veggie shish-ka-bobs and ranch dressing? Mozzarella, heirloom tomato and basil mini-pizzas? Small watermelons, rhubarb pie, make-your-own salad bar? Particularly if the child has participated in growing the garden, they will really enjoy showing it off to their friends... and you are introducing more families to gardening!

Salad Days
Give your child a special basket and pair of scissors, and let them be responsible for snipping fresh lettuce, herbs and edible flowers from your garden for the evening side salad. Some guidance may prove necessary at first.

Edible Flowers
Eating beautiful, colorful flowers is a wonderful experience, especially for children. Petals from nasturtiums, pansies, chive flowers, carnations, calendula, chrysanthemums, English daisies, fuchsia, hibiscus, lilac, sunflowers, snapdragons and violets are all edible, as well as many others. When in doubt, check it out! Flowers make salad so much more interesting.

Harvest Photo Cards
Christmas photo cards are quite popular these days, but have you ever considered trying other seasons as well? Working particularly well for extended family, photo collage cards are quite inexpensive and a great way to stay in touch. Consider taking photos of your little gardeners during the Spring, Summer and harvest-time, then put them together in a collage and print enough to mail and a couple to save.

Eating Flowers

Salmon River Pumpkin Story (a winter squash) going from seed to stomach back to seed

Warning, these squash plants do not grow well during extended rainy seasons, too much moisture. Instead plant during the dry season.

Day 7 to 14 -- Germination

Day 1 -- Planting one seed

Day 90 -- Seed Saving & Ripe Harvest
200 seeds from one pumpkin

Success in Africa

Day 75 -- Green Harvest, Cooking & Early Eating

Day 40 -- First Flowers

Day 45 -- Pollination

Day 60 -- Pumpkins

One Great Big Smile & One BIG Potato
She will grow up healthy and strong,
eating from food growing in rich compost.
By the way ... 4 of these potatoes = 10 pounds

Gone Wild - Growing Healthy Plants Without Money

Look at these photos. We do have a choice. The flowers on the left are growing in one of our garden beds that we do not plant anymore. They have "Gone Wild" - meaning we do not plant, hoe, till, or cultivate. We just let the bed "Go Wild" with no fertilizing, no spraying, and no more hard work. This bed grows flowers each year all by itself. The flowers self-seed, and the soils full of compost and earthworms self-fertilize. We are trying hard to copy nature's processes in our garden. We do mulch in the fall just as nature does.

Another great example of "Gone Wild" happens each year with one of our neighbors who has a fair-sized lettuce patch. She lets a few of her lettuce plants go to seed, and the following spring another new patch of lettuce plants start growing as soon as the soil warms up. No work, she just stands back and lets nature do its thing.

We hope that you, too, can spend less money, do less work and enjoy growing and eating healthy food. That is what "Gardening for Life - No Money Required" is all about.

You can do this too!

Make your own soil **Plant in 100% compost** **Eat healthy**

God bless you & your gardens

Success in the Garden is Just the Beginning
So Let's - Enjoy the Journey